AF213038

MICROTECHNOLOGY AND MEMS

The photo shows a scanning electron micrograph of a shape memory microactuator. The microactuator has been developed by Dr. Klaus Skrobanek and the author for control of the deflection of a membrane. It consists of eight double beams, whose lateral shape has been designed for optimum use of the shape memory effect upon bending actuation. Fabrication has been done by laser cutting of a cold-rolled thin plate of NiTi.

M. Kohl

Shape Memory Microactuators

With 149 Figures

Springer

PD Dr. rer. nat. habil. Manfred Kohl
Forschungszentrum Karlsruhe
Institut für Mikrostrukturtechnik
Postfach 3640
76021 Karlsruhe, Germany
E-mail: manfred.kohl@imt.fzk.de

Series Editors:

Professor Dr. H. Baltes
ETH Zürich, Physical Electronics Laboratory
ETH Hoenggerberg, HPT-H6, 8093 Zürich, Switzerland

Professor Dr. Hiroyuki Fujita
University of Tokyo, Institute of Industrial Science
4-6-1 Komaba, Meguro-ku, Tokyo 153-8505, Japan

Professor Dr. Dorian Liepmann
University of California, Department of Bioengineering
466 Evans Hall, #1762, Berkeley, CA 94720-1762, USA

ISSN 1439-6599
ISBN 978-3-642-05837-0 ISBN 978-3-662-09875-2 (eBook)
DOI 10.1007/978-3-662-09875-2

Library of Congress Control Number: 2004102315

This work is subject to copyright. All rights are reserved, whether the whole or part of the material is concerned, specifically the rights of translation, reprinting, reuse of illustrations, recitation, broadcasting, reproduction on microfilm or in any other way, and storage in data banks. Duplication of this publication or parts thereof is permitted only under the provisions of the German Copyright Law of September 9, 1965, in its current version, and permission for use must always be obtained from Springer-Verlag Berlin Heidelberg GmbH. Violations are liable for prosecution under the German Copyright Law.

springeronline.com

© Springer-Verlag Berlin Heidelberg 2004
Originally published by Springer-Verlag Berlin Heidelberg New York in 2004
Softcover reprint of the hardcover 1st edition 2004

The use of general descriptive names, registered names, trademarks, etc. in this publication does not imply, even in the absence of a specific statement, that such names are exempt from the relevant protective laws and regulations and therefore free for general use.

Typesetting by the author
Cover concept: eStudio Calamar Steinen
Cover production: *design & production* GmbH, Heidelberg

Printed on acid-free paper 10884414 57/3141/ba - 5 4 3 2 1 0

Contents

Preface

Shape memory alloys exhibit the fascinating property of shape recovery, which allows them to generate large work outputs particularly in small dimensions. Their large potential for applications as "smart materials" is due to a number of additional unique features like direct electrical control by small voltages, multifunctional performance, and biocompatibility. Upon miniaturization, intrinsic disadvantages such as long time constants loose significance. Thus, shape memory alloys are virtually predestined for applications in microsystems technology, which is presently rapidly evolving.

This book combines fundamental aspects of materials science and micromechanics with selected design examples of demonstrators needed for a microsystems engineer to develop shape memory microactuators. While the fundamentals are written in a textbook manner, the development of the demonstrators is described from the idea to the prototype in comprehensive breadth. For evaluation of achieved specifications, the book also contains a review of comparable microactuator developments based on alternative actuation principles.

The purpose of this book is to disseminate recent progress and underlying fundamentals of shape memory microactuators and to provide practicable approaches for future developments of processes and products. The main intended audience are engineers and engineering students from the materials, microsystems, and applications communities. The reader not familiar with shape memory alloys will find all the information required for getting started. I hope that this book will help to increase the awareness for this new field. Any comments or suggestions regarding this work are welcome.

The book originated from several research projects at Forschungszentrum Karlsruhe and a lecture on microactuators at the University of Karlsruhe. By and by, a specific approach to designing, modeling and fabricating shape memory microactuators has been developed based on monolithic components of micromachined foils or thin films of shape memory alloys. Using this approach, several prototypes of microvalves, linear actuators, and microgrippers have been developed that serve as demonstrators in this book.

Various functional units may be realized in a monolithic shape memory component, which enables cost-effective fabrication by a single process step. The variety of structures and functions requires, however, a procedure to design optimized lateral shapes, which enable an optimum use of the shape memory effect. For this purpose, mechanical and thermal optimization criteria are introduced and their implementation is shown for the demonstrators.

Modeling realistic components requires either an approximate analytical or a sufficiently meshed numerical representation of the governing equations. The use of analytic models provides invaluable design insights. When dealing with an entire microactuator, the use of a macromodel capturing the essential physical behavior appears most appropriate. In this book, analytic and numeric macromodels are presented for design simulation and verification. As the electrical, thermal, and mechanical properties of shape memory microactuators depend strongly on each other, a finite element routine for coupled multilevel simulation is introduced to achieve a quantitative description.

Shape memory microactuators may be fabricated by monolithic integration following e.g. the procedure of surface micromachining in silicon technology or by hybrid integration of shape memory microstructures fabricated separately from the final microsystem. The book covers selected examples of both approaches. The latter approach allows integration of shape memory microactuators in polymer-based microsystems, which are a main focus at Forschungszentrum Karlsruhe. Hybrid integration also offers a larger flexibility with respect to the choice of processes such as micromachining, heat treatment, etc. and related parameters. Consequently, the demonstrators in this book have been fabricated by hybrid integration.

I am indebted to Prof. V. Saile, Prof. H. Janocha, and Prof. K.H. Zum Gahr for critically reading the manuscript, which was first written and published in German as a professorial dissertation. Parts of the English translations have been done by Mrs. Schröder from the translation bureau of Forschungszentrum Karlsruhe. The presented developments of shape memory microactuators at Forschungszentrum Karlsruhe rely on the support and input of many individuals. Special thanks go to my colleague Dr. Berthold Krevet, an expert in finite element modeling, with whom I spent many hours discussing various issues of engineering, physics, and more. Berthold contributed significantly to the implementation of numeric macromodels for the demonstrators presented in this book and provided many helpful comments. I would like to thank my colleague Siegfried Dürr for his competent technical support. His input in solving technological problems of assembly and interconnection has been of immense help. I also thank my past and present students for their contributions. Their excitement about their work and the good spirit in the research group have been essential for the various achievements. The shape memory microactuators have been fabricated from high quality SMA foils and thin films provided from several sources. I am particularly thankful to Prof. Shuichi Miyazaki, Dr. Eckhard Quandt, Dr. Bernhard Winzek, Prof. Toshiyuki Takagi, and Kiyoshi Yamauchi for the very fruitful collaboration over many years. In addition, Prof. David Allen, Dr. Andreas Schüßler, Dr. Wilhelm Pfleging, and Heino Besser deserve my sincere thanks for their support in micromachining. Last but not least, I would like to thank my wife Birgit for putting up with my preoccupation and long working hours.

Karlsruhe, February 2004 Manfred Kohl

Glossary of Notations

This glossary lists the abbreviations and symbols used in the book, a brief definition of their meaning, and typical units. Some symbols have different meanings in separate contexts, matching common notation conventions.

Notation	Description	Units
a	Lattice parameter	m
a	Radius of bridge in SMA microvalve	m
a_B	Length of a bridge segment in SMA microvalve	m
a^A	Parameter for the kinetic behavior of phase transformation	K^{-1}
a^M	Parameter for the kinetic behavior of phase transformation	K^{-1}
A	Analog	
A	Ampère	
A	Austenite	
A	Area	m^2
A_s	Austenite start transformation temperature	K
A_p	Austenite peak transformation temperature	K
A_f	Austenite finish transformation temperature	K
Al	Aluminium	
At	Atom	
Au	Gold	
ASP	Analog signal processing unit	
b	Lattice parameter	m
b_B	Width of a bridge segment in SMA microvalve	m
b^A	Parameter for the kinetic behavior of phase transformation	Pa^{-1}
b^M	Parameter for the kinetic behavior of phase transformation	Pa^{-1}
B	Magnetic field strength	Tesla
B	Operator of lattice deformation	
c	Lattice parameter	m

Notation	Description	Units
c	Stress rate	PaK^{-1}
$c^{M,A}$	Stress rate in martensitic, austenitic condition	PaK^{-1}
c_p	Specific heat capacity at constant pressure	$Jkg^{-1}K^{-1}$
conv	Convective	
cond	Conductive	
const	Constant	
C	Capacity	AsV^{-1}
C	Fluidic friction coefficient	
C	Carbon	
Cr	Chromium	
Cu	Copper	
Cal	Calorie	
CAD	Computer-aided design	
CAO	Computer-aided optimization	
CMOS	Complementary metal oxide semiconductor	
d	Thickness of SMA thin plate, foil, or thin film	m
d	Vector or lattice invariant shear	
d_I	Unit vector	
$d_{k\lambda}$	Piezoelectric coefficient	CN^{-1}
dB	Decibel	
D_h	Hydraulic diameter (4·cross-sectional area/perimeter)	m
D	Digital	
DC	Direct current	
DSC	Differential scanning calorimetry	
el	Elastic	
exp	Experimental	
E	Electrical field strength	Vm^{-1}
E	E-modulus (Young's modulus)	Pa
E^*	E-modulus in computer-aided optimization	Pa
$\overline{E}$	Effective E-modulus	Pa
$\overline{E}^R$	Effective E-modulus in R-phase condition	Pa
$E^{M,A}$	E-modulus in martensitic, austenitic condition	Pa
E	Energy	J
EQ	Point of force equilibrium	

Notation	Description	Units
F	Force	N
F^R	Force in R-phase condition	N
$F^{M,A}$	Force in martensitic, austenitic condition	N
F_c	Closing force	N
F_G	Gravitational force	N
F_{Δ_p}	Pressure force	N
F^{SE}	Spring force	N
FK	Spring constant	Nm^{-1}
F	Frequency	Hz
Fe	Iron	
FEM	Finite element method	
g	Gravitational constant	ms^{-2}
g	Gram	
G	Gibbs potential	Jm^{-3}
h	Beam thickness	m
h	Latent heat of phase transformation	$Jkg^{-1}K^{-1}$
Hf	Hafnium	
Hz	Hertz	
i	Iteration step	
i	Coordinate index	
I	(3x3)-identity matrix	
I	Current	A
I_y	Axial surface moment of second degree	m^4
I	Light intensity	Wm^{-2}
I	Information	
IC	Integrated circuit	
j	Current density	Am^{-2}
J	Joule	
k_0	Fit parameter for convective cooling power	
$[k]$	Tangential stiffness matrix for each element	
kg	Kilogram	
$[K]$	Global tangential stiffness matrix	
K	Heat transfer coefficient	$Wm^{-2}K^{-1}$
$\overline{K}$	Effective heat transfer coefficient	$Wm^{-2}K^{-1}$

Notation	Description	Units
K	Kelvin	
l	Length	m
L	Inductance	VsA^{-1}
L	Gap length of a valve seat in flow direction	m
LED	Light emitting diode	
LIGA	Lithography-electroplating-molding (In German: Lithographie-Galvanik-Abformung)	
m	Meter	
m_l	Amount of shape strain	
m	Mass	kg
ms	Millisecond	
max	Maximum	
M	Magnetization	Am^{-1}
M_b	Bending moment	Nm
M	Martensite	
$M^{+,-}$	Martensite variants	
M_s	Martensite start temperature	K
M_p	Martensite peak temperature	K
M_f	Martensite finish temperature	K
M_d	Temperature limit of stress-induced martensite formation	K
Mn	Manganese	
MST	Microsystems technology	
MOCVD	Metalorganic chemical vapor deposition	
n	number of load steps in finite element simulation	
n_B	Number of segments of bridge	
n.c.	Normally closed	
n.o.	Normally open	
N	Newton	
N	Number of load steps in multilevel simulation	
N	Number of turns of a coil	
N	Magnetic north pole	
Nb	Niobium	
Nd	Neodymium	
N_2	Nitrogen	
Ni	Nickel	

Notation	Description	Units
o	Oriented	
O	Oxygen	
p_1	Unit vector perpendicular to habitus plane	
p	Pressure	Pa
p_{in}	Inlet pressure	Pa
p_{out}	Outlet pressure	Pa
p_i	Internal pressure	Pa
p_e	External pressure	Pa
pe	Pseudoelastic	
P_1	Shape strain matrix	
P_2	Operator of lattice invariant shear	
P	Power	W
P	Polarization	Asm^{-2}
$\{P\}$	External forces	N
P_{el}	Electrical heating power	W
P_{conv}	Convective power	W
P_{cond}	Conductive power	W
Pa	Pascal	
Pb	Lead	
Pd	Palladium	
Pt	Platinum	
PID	Proportional integral differential	
PSU	Polysulphone	
PZT	Lead zirconium titanate	
Q	Electrostatic charge	As
Q	Volumetric flow rate	Sccm
Q	Heat energy	J
Q	Heating power	W
r	Scaling variable	m
rd	Parallel to rolling direction	
R	Radius of curvature	m
R	Electrical resistance	Ohm
R	Operator of lattice rotation	
R	Rhombohedral phase	
R_s	Start temperature of R-phase	K

Notation	Description	Units
R_p	Peak temperature of R-phase	K
R_f	Finish temperature of R-phase	K
R&D	Research and development	
Re	Reynolds number	
Re_t	Critical Reynolds number	
Ref	Reference	
s	Second	
s	Displacement	m
s^{SE}	Displacement of biasing spring	m
s_{tot}	Total predeflection (linear actuator)	m
sa	Self-accommodated	
S	Entropy	$Jm^{-3}K^{-1}$
S_1	Displacement of linear actuator of SMA microgripper	m
S_2	Gripping stroke	m
S	Magnetic south pole	
S	Substance	
Sccm	Standard cubic centimeter	
Si	Silicon	
SIM	Stress-induced martensite	
SMA	Shape memory alloy	
t	Time	s
td	Transverse direction	
tr	Transformation	
T	Temperature	K
T_E	Environmental temperature	K
T_0	Equilibrium temperature	K
Ti	Titanium	
$\{u\}$	Element displacements	m
U	Work	Nm
U	Intrinsic energy density	Jm^{-3}
$\{U\}$	Global displacements	m
UV	Ultraviolet	
V	Volume	m^3

Notation	Description	Units
V	Voltage	V
V	Volt	
w	Beam width	m
W	Watt	
WLR	Wechsler-Lieberman-Read	
YAG	Yttrium-aluminium-garnet	
x	Space coordinate	m
x_R	Fraction of R-phase	
y	Space coordinate	m
z	Space coordinate	m
$z^{A,M}$	Displacement in austenitic, martensitic condition	m
z^R	Displacement in R-phase condition	m
Zn	Zinc	
Zr	Zirconium	
α	Angle of displacement	Degree
α	Lattice parameter	Degree
α	Thermal expansion coefficient	K^{-1}
α^*	Thermal expansion coefficient in CAO optimization	K^{-1}
α'	Thermal coefficient	$Nm^{-2}K^{-1}$
β	Lattice parameter	Degree
γ	Lattice parameter	Degree
γ	Shear angle	Degree
γ_G	Shear limit	Degree
Δz	Stroke of SMA actuator in z-direction	m
ΔH	Transformation enthalpy	Jkg^{-1}
ΔT_{S1}	Supercooling	K
ΔT_{S2}	Superheating	K
ε	Strain	
ε_A	Strain of reverse transformation (to austenite)	
ε_p	Plastic strain	
$\varepsilon^{A,M}$	Maximum strain in austenitic, martensitic condition	
ε^R	Maximum strain in R-phase condition	

Notation	Description	Units
$\{\varepsilon\}$	Total strain matrix (sum of elastic and transformation strain)	
$\bar{\varepsilon}$	Effective strain	
$\bar{\varepsilon}^{A,M}$	Maximum effective strain in austenitic, martensitic condition	
$\bar{\varepsilon}^{R}$	Maximum effective strain in R-phase condition	
$\bar{\varepsilon}^{tr}$	Effective transformation strain	
$\{\varepsilon\}^{el}$	Elastic strain matrix	
$\{\varepsilon\}^{tr}$	Transformation strain matrix	
λ	Heat conductivity	$Wm^{-1}K^{-1}$
μ	Permeability	
μm	Micrometer	
μ_0	Permeability of free space	$VsA^{-1}m^{-1}$
μR	Microprocessor	
ρ	Density	kgm^{-3}
ρ	Specific electrical resistance	Ohm·m
$\{\rho^{-1}\}$	Electrical conductivity matrix	
σ	Tensile stress	Nm^{-2}
σ_S	Edge stress	Nm^{-2}
σ_{limit}	Stress limit	Nm^{-2}
$\sigma^{A,M}$	Maximum tensile stress in austenitic, martensitic condition	Nm^{-2}
σ^{R}	Maximum tensile stress in R-phase condition	Nm^{-2}
$\{\sigma\}$	Total stress matrix (sum of austenitic and martensitic, respectively R-phase contributions)	
$\{\sigma\}^{A,M}$	Stress matrix of austenitic, martensitic condition	
$\{\sigma\}^{R}$	Stress matrix of R-phase condition	
$\bar{\sigma}$	Effective stress	Nm^{-2}
$\bar{\sigma}^{A,M}$	Maximum von Mises stress in austenitic, martensitic condition	Nm^{-2}
$\bar{\sigma}^{R}$	Maximum von Mises stress in R-phase condition	Nm^{-2}
τ	Shear stress	Nm^{-2}
τ_G	Limit of shear stress	Nm^{-2}
τ_h	Heating time	s

Notation	Description	Units
τ_c	Cooling time	s
τ_s	Closing time	s
τ_o	Opening time	s
τ_{sw}	Switching time	s
τ_{act}	Duration of heating pulse	s
τ_{rel}	Time interval between subsequent heating pulses	s
ξ	Martensitic phase fraction	
ζ	Fluidic parameter	
Φ	Electrical potential	V
1-D	One-dimensional	
2-D	Two-dimensional	
3-D	Three-dimensional	
1W	One way	

1 Introduction and Overview

The definition of terms is the beginning of wisdom.

–Sokrates

Actuators are essential components for the control of processes. The areas of application reach from production equipment and household appliances of macroscopic size, for instance, to cardiac pacemakers and bubble-jet printers, which can be regarded as microsystems products.

Until today the term "actuator" was neither determined in standards nor in regulations. Following the habitual language use, an actuator may be considered a link between the information processing part of a control system and a process [1, 2]. As shown schematically in Fig. 1.1, an actuator affects processes in its environment, through which, generally, any flow of information, energy or substance is changed. In many cases, the input of an actuator is controlled electrically. In principle, however, non-electric e.g. optical or fluidic control signals are also possible. Intelligent functions are generated for example by recording process parameters with sensors and, by further processing, generating suitable control signals in order to change the process parameters in accordance with predetermined objectives.

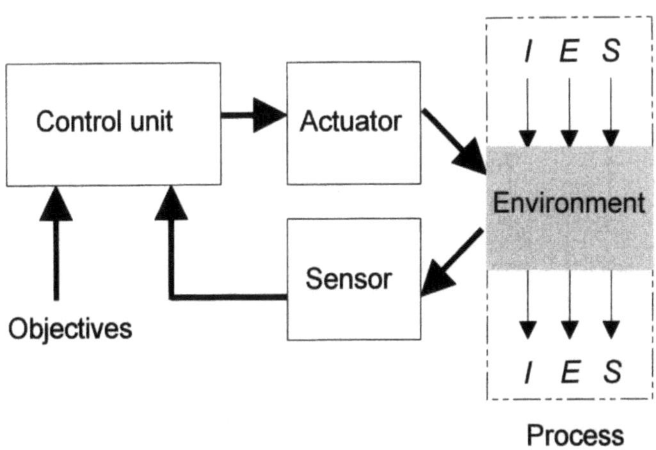

Fig. 1.1. Control system. The abbreviations *I*, *E*, and *S* denote information, energy, and substance

Concerning the term "microactuator" no habitual language use has prevailed so far, as is generally the case in microsystems technology. Constitutive features of microactuators are that critical dimensions of function determining structures are in the millimeter range or that microtechnologies are essential for their fabrication.

The development of microactuators is still a young field of activity, whose beginnings date back to the late seventies, see, e.g., [3, 4]. In the nineties, microactuator development evolved very dynamically. The continously emerging demand on microactuators can be considered a direct consequence of the progress in the field of microelectronics. By reducing dimensions, power consumption, and costs, decentralized control systems gain increased importance, where parts of information processing are moved out of the central control unit to the process periphery. As sketched in Fig. 1.2, this concept leads to the development of "intelligent" sensors and actuators, which are equipped, in particular, with microprocessors and communicate via bus systems.

A further development is the integration of several microactuators and sensors to microsystems, by which novel functions are enabled in small dimensions and an overall enhanced capability is achieved.

The diversity of materials and functions imply that the field of microactuators develops toward a multidisciplinary field of research. This trend is enhanced by the fact that the systems aspect gains a special importance due to the smallness and proximity of microcomponents. For example, the properties of microactuators may be strongly influenced by heat generation or by electromagnetic radiation of neighbouring circuits. Therefore, microactuators generally cannot be considered isolated components. Instead, non-negligible coupling and interference may lead to very complicated system properties, which must already be taken into account in the actuator design.

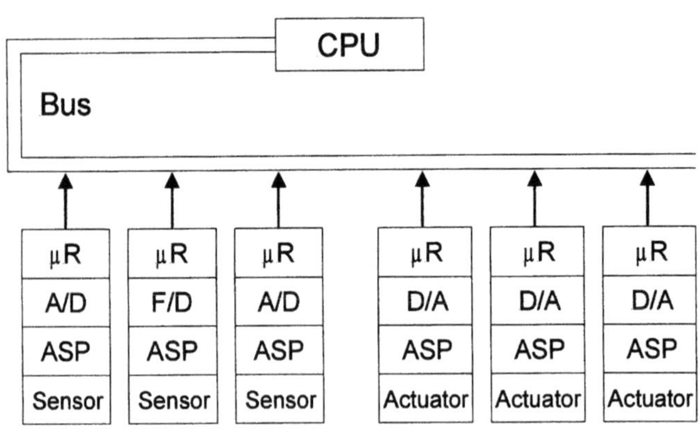

Fig. 1.2. Decentralized control system. The abbreviations *CPU*, *μR*, *A/D*, *F/D*, *D/A*, and *ASP* denote the central processing unit, microprocessor, analog/digital, frequency/digital, digital/analog transducer, and analog signal processing unit, respectively

Generally, the development of microactuators is associated with high requirements for design methods, materials, and technologies, so that cost-effective solutions are rather an exception and a broad market launch has not yet been possible. Thus, only a few commercially available microactuator systems exist so far.

One impressive example is the bubble-jet printhead, which reached a production volume over 100 million in 1996, corresponding to a market volume above $3 billion [5]. Besides fluidic microactuators, mechanical microswitches and micropositioners are gaining importance. Typical examples are optical mirror switches for fiber-optical telecommunication or microrelais for the replacement of present electromechanical relais.

Currently, two approaches are being pursued in the development of microactuators. One trend is the innovative transfer of conventional, in macrodimensions approved principles to microdimensions by applying microtechnologies. The electromotor is one example, which could be miniaturized to a micromotor of 1.9 mm diameter [6]. A second trend is the introduction of so-called "new" actuation principles [7], e.g., of the inverse piezoeffect or the shape memory effect, which bear promising application potentials. The term "new" in this context is not related to the time of discovery, but emphasizes that the actuation principles have gained an appreciable importance only recently due to substantial progress in their technological implementation.

Among the presently known actuation principles, the shape memory effect shows the highest energy densities, which are on the order of 10 J/cm^3 for NiTi alloys [8]. For this reason, the shape memory effect appears particularly interesting for applications where, despite small dimensions, large forces and displacements are required. Shape memory alloys (SMAs) are attributed to so-called "smart" materials, since they may fulfill, aside from actuation functions, temperature sensing, electrical, or structural functions and thus enable simple and compact designs with multifunctional features [9]. Even the characteristic disadvantages of macroscopic SMA actuators of low thermodynamic efficiency and low heat transfer rates become less important for decreasing dimensions. Thus, for dimensions in the micrometer range, cycle times of a few milliseconds are possible [10]. Due to these reasons, SMA actuators are virtually predestined for use in microsystems.

First attempts at the development of SMA microactuators have addressed the miniaturization of macroscopic designs consisting of SMA wires or plates [11]. This approach, however, does not lead to technologically and economically workable solutions. The realization of SMA microactuators requires novel approaches following the methods of microelectronics. In recent years, the development of SMA thin films through sputtering methods [12–18] and of SMA foils by extending rolling technology [19, 20] has created a technological basis for the development of SMA microactuators. Thus, the field of SMA microactuators represents a field of activity, which comprises, besides the domain of materials science and engineering, various other aspects such as design, modeling, simulation, optimization, fabrication technologies (micromaching, packaging, and interconnection technologies), and control issues.

The following chapter addresses the present development stage of microactuators by means of microvalves and linear actuators. These examples allow the demonstration of representative properties and problems, which characterize the current state of the art of microactuators. The SMA microvalves and linear actuators presented in Chaps. 5 and 6 will be evaluated with respect to this background.

Chap. 3 contains the basics of shape memory effects, which are relevant for the development of SMA microactuators. Besides an introduction to cristallography and thermodynamics, the particular properties of NiTi-based alloys will be presented.

In Chap. 4, the most relevant principles for the development of SMA microactuators will be introduced. Central topics are the simulation and the design optimization of SMA components. Furthermore, problems of SMA actuator design and fabrication technologies will be addressed.

The subsequent Chaps. 5 and 6 present selected design examples of demonstrators developed at Forschungszentrum Karlsruhe. Chap. 5 introduces various variants of SMA microvalves, which differ with respect to their function (normally open/normally closed), structure, and fabrication. Chap. 6 presents SMA linear actuators and their further development to optical microswitches and microgrippers.

Finally, a summary is given in Chap. 7.

2 Microactuators

And it turns out that all of the information that man has carefully accumulated in all the books in the world can be written in this form in a cube of material one two-hundredth of an inch wide–which is the barest piece of dust that can be made out by the human eye.

–R.P. Feynman in *There's Plenty of Room at the Bottom*

Presently, considerable efforts are undertaken worldwide to develop microsystems, which may comprise various functional units, including mechanical, optical, and fluidic components and subsystems. Figure 2.1 gives an overview of application fields that are of relevance to microsystems technology [21].

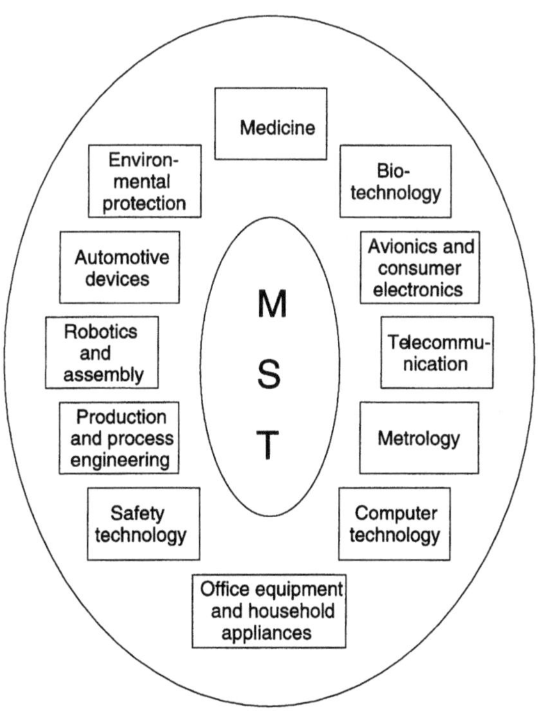

Fig. 2.1. Microsystems applications. The acronym *MST* stands for microsystems technology [21]

Frequently required actuation functions are the handling of small amounts of fluid, e.g., in medical, biological, and chemical engineering, as well as the execution of switching or positioning functions in information technology, safety technology, and robotics, for instance. In the present chapter, particular attention will be paid to both of these areas. Major microactuators for handling small fluid volumes are microvalves. They are required, for example, to control pressure gradients and flow rates in a fluid channel or to switch fluid flows between various fluid channels. Major microactuators fulfilling switching or positioning functions are linear actuators.

Following a comparative discussion of various currently available actuation principles, the present state of the art will be described for both types of actuators.

2.1 Actuation Principles

Functional properties of actuators are determined above all by the underlying actuation principles. Consequently, the selection of actuation principles is of crucial importance. Table 2.1 gives an overview of various actuation principles that are currently relevant to the development of microactuators. In the last column, the work densities achieved with the materials and technologies available today are compared. The forces and displacements that can actually be reached for a certain component size depend on the scaling behavior, actuator design, and technical limits. Therefore, the total dimensions of an actuator determine which actuation principle is best suited for a given application. The scaling behavior of the forces is derived from the force laws in Table 2.1 by introducing a scaling variable *r*, which describes the component size in any spatial direction. In analogy, the scaling behavior of work and power density can be determined. The results are presented in Table 2.2 for various actuation principles.

Fig. 2.2 gives an overview of the technically achievable power densities, normalized to the actuator mass ranging from 10^{-4} to 10^{7} g. Most data have been taken from a Japanese study [22]. The microactuator data have been obtained for SMA linear actuators and SMA valve actuators described in this book, as well as for electrorheological microvalves presented in [23]. Maximum power densities on the order of 10 W/g are reached in gas turbines and hydraulic motors with actuator masses ranging from 10^{4} to 10^{6} g. With decreasing actuator mass, a tendency toward smaller power densities can be observed. In the mesoscopic range between 10^{-1} and 10 g, piezoactuators have already found their way into industrial mass-production applications. Below 10^{-1} g, the regime of microactuators follows. Here, so-called "new actuators", such as electrorheological or SMA actuators show promising specifications.

Apart from work and power density, there are numerous other criteria for evaluating actuators (time constants, accuracy, efficiency, service life, etc.). They will be dealt with in some detail using microvalves and linear actuators as examples.

Table 2.1. Overview of frequently applied actuation principles. The symbols are specified in the Glossary of Notations

Principle	Scheme	Force F_i	Maximum work density U/V
Electrostatic: - plate attraction - plate alignment	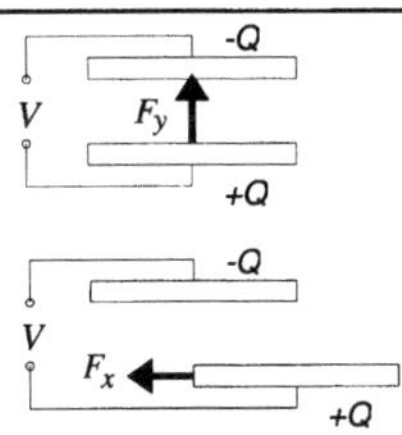	$F_i = \frac{\partial}{\partial i}(\frac{1}{2}CV^2)$	10^5 J/m³ (assumption: $E_{max} = 1.5 \cdot 10^8$ V/m)
Magnetic: - electrodynamic - reluctance - electromagnetic	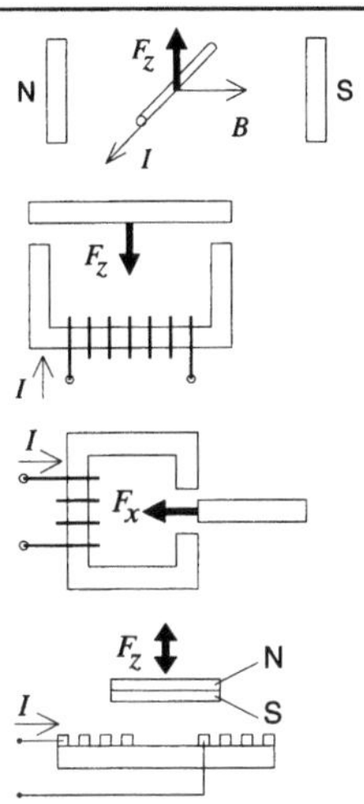	$dF_i = \frac{\mu_0}{4\pi} I(d\vec{l} \times d\vec{B})_i$ $F_i = \frac{\partial}{\partial i}(\frac{1}{2}LI^2)$ $F_i = \frac{\partial}{\partial i}(\vec{M}\vec{B})$	$4 \cdot 10^5$ J/m³ (assumption: B_{max} = 1 Tesla)
Fluidic: - piston displacement - membrane deflection	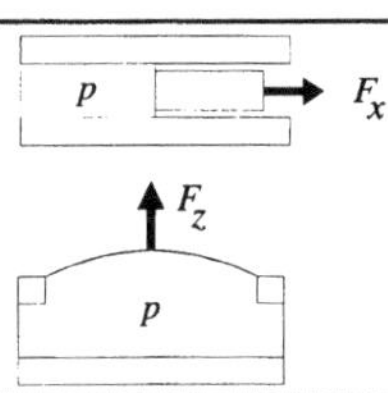	$dF_i = p\,dA_i$	10^6 J/m³ (assumption: $p_{max} = 10^6$ Pa)
Inverse piezoeffect: - displacement principle - bending principle	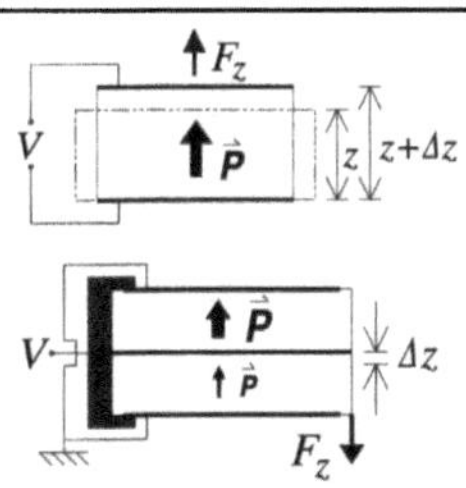	$dF_i = AE_{ik}d_{k\lambda}\,dE_k$	$1.2 \cdot 10^4$ J/m³ (assumption: $E_{max} = 10^6$ V/m)
Shape memory effect: - tensile effect	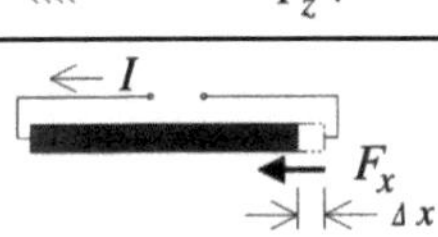	$F_x = \sigma_x A$	10^7 J/m³ (assumption: ε_{max} = 4%)

Table 2.2. Scaling behavior of frequently applied principles in microactuator technology. The underlying assumptions are explained in the text. The scaling variable r describes the component size in any spatial direction

Principle	Assumption	Force	Work density	Power density
Electrostatic	$E \propto r^0$	$F \propto r^2$	$U/V \propto r^0$	$P/V \propto r^{-1}$
Reluctance	$J \propto r^0$	$F \propto r^4$	$U/V \propto r^2$	$P/V \propto r^2$
	$J \propto r^{-1}$	$F \propto r^2$	$U/V \propto r^0$	$P/V \propto r^{-1}$
Fluidic	$p \propto r^0$	$F \propto r^2$	$U/V \propto r^0$	$P/V \propto r^{-1}$
Inverse piezoeffect	$E \propto r^0$	$F \propto r^2$	$U/V \propto r^0$	$P/V \propto r^{-1}$
Shape memory effect	$J \propto r^0$	$F \propto r^2$	$U/V \propto r^0$	$P/V \propto r^0$
	$J \propto r^{-1}$	$F \propto r^2$	$U/V \propto r^0$	$P/V \propto r^{-2}$

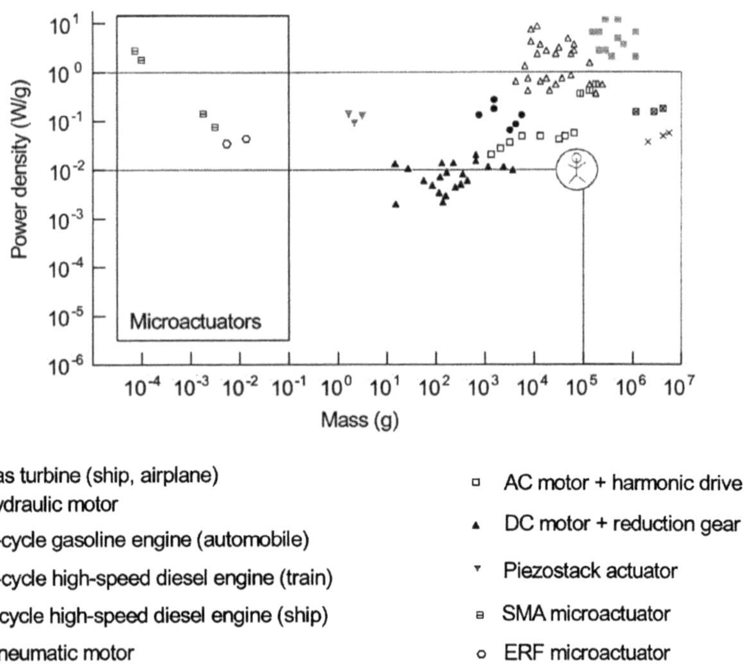

Fig. 2.2. Achievable power densities of various actuators in macro and micro dimensions, normalized to the actuator mass. The typical power density of man is given for comparison

2.1.1 Electrostatic Principles

Electrostatic actuators make use of Coulomb's attraction force between oppositely charged bodies. In the most simple case, two charged plates oriented in parallel are available, as shown in Table 2.1. Apart from the force component F_z that acts vertically between the plates, lateral-plate offset additionally causes a force F_x acting in the lateral direction. In case of a nonparallel geometry or flexible capacitor elements (e.g., bending or membrane structures), both force components are used at the same time. In other arrangements, the forces between bodies of different dielectric constants are used.

The energy in the capacitor is proportional to the capacity of the actuator C and, hence, indirectly proportional to the distance of the charged plates. As a consequence, plate distance is an important design parameter apart from the plate surface area. Plate distance should be chosen as small as possible in order to reach satisfactory forces. Here, the advantages of microtechnology play a decisive role.

Today, it is possible to reach high electric field strengths in thin insulation layers with small geometric tolerances. For small gap dimensions below 10 µm, the maximum allowable field strength E_{max} deviates from macroscopic limits due to different physical mechanisms causing electrical breakdown [24].

As the vertical force component F_z drops very rapidly with increasing plate distance, the lateral force F_x is crucial to generating large displacements. Interestingly, the force component F_x is independent of the plate overlap without the plate length contributing. Therefore, it is more reasonable to have many plate poles switched in parallel, whose forces are added. This leads to electrostatic linear and rotation actuators with numerous capacitor elements arranged on a large area. The maximum available work density of these actuators results from the maximum available actuator area and the technically achievable geometrical tolerances of the capacitor gaps and heights in this area. Consequently, high aspect ratios are particularly desirable. Using the LIGA process, high-performance electrostatic linear actuators have been fabricated, the specifications of which are given in Section 2.3.

As evident from Table 2.2, electrostatic forces possess a very favorable scaling behavior, in particular when approximating the scaling behavior of the field strength by $E \propto r^0$, which seems reasonable for gap dimensions larger than 10 µm. As a result, high work and power densities can be achieved even for electrostatic microactuators of small size.

2.1.2 Magnetic Principles

The magnetic principles of force and movement generation can be classified in electrodynamic, reluctance, and electromagnetic principles. Electrodynamic microactuators make use of the Lorentz force on a conductor passed by a current in a magnetic field. So far, such actuators have hardly been noticed due to their difficult-to-achieve three-dimensional geometry. Contrary to this, the reluctance principle is the most frequently applied magnetic actuation principle in microtechnology. Microtechnical implementation requires an acceptable effort, since only a single field source and no hard magnetic materials are needed. As shown in Table 2.1, the electromagnetic principle requires two field sources, e.g., a coil with or without a core and a permanent magnet of magnetization $\vec{M}$. Other magnetic actuation principles not mentioned here are magnetostriction and the Meissner effect [2].

To generate high energy densities, high magnetic field strengths are required. In many cases, it is therefore favorable to conduct the field in a core of high permeability. In reluctance actuators, the energy converted is determined by the active volume in the air gap, which is made up of the air gap surface area and the air gap distance. Both are major design parameters that are usually optimized such that the magnetic reluctance in the air gap equals the reluctance in the core [25]. High forces in particular require small air gaps, which can be fabricated microtechnically with high precision. A major advantage as compared to electrostatic principles lies in the fact that no electric discharges occur. The maximum possible work density U/V is determined by the maximum possible current density in the field-generating coil, which is limited by the melting temperature and electromigration, or by the saturation field strength B_{max} in the magnetic core. Today, saturation field strengths B_{max} of 1 Tesla can be reached in microdimensions. The corresponding energy density U/V amounts to 4×10^5 J/m^3 and, hence, is of the same order of magnitude as that of electrostatic microactuators.

Manufacture of the field sources today represents a considerable challenge to available technologies. Planar microcoils require high numbers of windings and, thus, large actuator surface areas together with high serial resistances. Integrated manufacture of three-dimensional coils requires a considerable effort. To conduct the flux, a core material of high permeability is needed to minimize the core dimensions. The relatively large space required by the field-generating and the field-conducting structures leads to a bad scaling behavior of the magnetic forces. Table 2.2 shows the scaling behavior of the magnetic force based on the example of the reluctance principle in the case of a dimension-independent current density ($J \propto r^0$) and for the case of an increasing current density due to improved cooling in small dimensions according to $J \propto r^{-1}$ [26]. At a dimension-independent current density, the work and power densities drop quadratically with decreasing component size. This is the most unfavorable scaling behavior of all actuation principles presented in Table 2.2.

2.1.3 Fluidic Principles

Table 2.1 presents two examples of fluidic principles used for microactuators. Fluidically driven micropistons allow high testing forces and large displacements to be reached in, e.g., micro tensile testing machines. The force-displacement characteristics exhibit a constant behavior [27, 28]. The principle of pressure-dependent membrane deflection is applied in a number of microvalve versions. Other fluidic arrangements are based on fluid dynamics principles or electrorheological and magnetorheological principles [2, 7].

By applying a pressure p, high work densities can be achieved, which are limited by the pressure stability of the chambers and feed lines. A pressure limit of p_{max} = 10 bar corresponds to a work density U/V of 10^6 J/m^3, for instance. However, such pressures cannot yet be generated by microtechnical supplies. Today, the pressure differences generated by micropumps are below 1 bar. To reach high work densities, an external pressure supply by a macroscopic pump is required.

Suitability of fluidic actuation principles for microtechnical applications is reflected by their favorable scaling behavior, see Table 2.2.

2.1.4 Inverse Piezoeffect

Piezoelectric actuators make use of the coupling of mechanical deformation and electric polarization in ferroelectric crystals, in crystals with a triad axis, or in certain polymers. A major class of materials is made up of $Pb(Zr_xTi_{1-x})O_3$ (PZT) crystals due to their high coupling factors of about 0.7 and piezoelectric coefficients $d_{k\lambda}$ of about 0.5 nm/V [7]. As an example, Table 2.1 shows two designs, a displacement and a bending transducer. Due to the inverse piezoeffect, the application of an electric voltage causes the thickness to increase in the polarization direction and to decrease in the vertical direction. The bending transducer is of particular importance for microactuators. It consists of either two piezoelectric layers ("bimorph") excited in transverse electric mode or a combination of a piezoelectric and a non-piezoelectric layer.

The work density U/V depends quadratically on the piezoelectric coefficient $d_{k\lambda}$ and the electric field strength E_k, with the maximum field strength E_{max} being limited by the depolarization field strength or the breakdown voltage. PZT actuators reach a work density of 1.2 x 10^4 J/m^3 at $E_{max} = 10^6$ V/m.

As in all solid-state actuators, forces are transmitted by stress. This results in the scaling behavior given in Table 2.2, which is proportional to r^2 as long as a dimension-independent electric field is assumed.

2.1.5 Other Principles

Previous developments also comprise a number of other actuation principles that are not listed in Tables 2.1 and 2.2. They include, for instance, thermal principles (thermal and thermopneumatic expansion, bimetal effect), which are partly used in microfluidics, as described in the following section. In contrast to these principles, the implementation of optical [29], electrochemical [30], or chemomechanical [31] principles is still at its beginning.

2.2 Microvalves

Microvalves are needed for precise control of fluid flows. They may be combined with microsensors, micropumps, and other fluidic components to fluidic microsystems. Thus, for instance, portable devices may be realized allowing low-cost analyses of high-quality fluids. Potential fields of application are in environmental monitoring, medical engineering, or industrial automation technology. At the moment, considerable efforts are being undertaken to design such fluidic microsystems and the corresponding microcomponents.

In general, one must distinguish between passive and active valves. While passive valves are actuated by the forces of the fluid to be controlled, active valves require an independent actuation mechanism. Passive microvalves have mainly been developed for flow control in micropumps and, in the most simple case, consist of a bending element and a sealing surface [32]. The major driving mechanisms implemented in active microvalves so far are based on bimetal [33–39] and thermopneumatic principles [38, 40–45] as well as on thermal buckling [46–51], piezoelectric [52–56], electrostatic [57–65], electromagnetic [59, 66–68], and shape memory principles [69–73]. Pneumatic microvalves will not be considered here, as they require an external pressure supply in most cases that cannot be replaced by available micropumps. A promising solution of this problem might be the development of an integrated electrochemical pressure supply [74, 75]. Table 2.3 gives a survey of the major specifications of selected active microvalves that reflect the current state of the art. For comparison, the specifications of a miniaturized magnetic valve of macroscopic design are given [76].

2.2.1 Electromagnetic Microvalves

In the first microvalve prototypes, miniaturized electromagnetic drives were applied with the magnetic field being generated by an external coil [4]. Minimum dimensions of such magnetic valves, which can be produced by high-precision engineering, are limited to about 10 x 10 mm^2 [77]. For further miniaturization, attempts were made to integrate the coils in the microvalve. However, realization of integrated coils with sufficient field strengths or field strength gradients still represents a considerable problem. For this reason, the pressure differences that can be controlled at the moment are relatively small, see Table 2.3. Another problem is the deposition of thin films with attractive magnetic properties. Advantages of magnetically driven microvalves comprise the compatibility with CMOS technology due to small impedances and the associated small operating voltages, as well as bidirectional triggering when operated with alternating current.

Fig. 2.3 shows a schematic cross-section of an electromagnetic microvalve [68]. The valve consists of two silicon chips. The actuation chip contains a membrane structure with electroplated gold coil, which is deflectable by electromagnetic force. The microvalve is completed by the encapsulation chip containing the valve seat, inlet and outlet, as well as a permanent magnet.

Table 2.3. Specifications of microvalves fabricated by silicon technology. The abbreviations n.o. and n.c. stand for normally open and normally closed. Due to the lack of data with regard to the switching times, the following reference values may be considered for orientation: * Electromagnetic microvalve: About 0.4 ms, ref. [59], ** thermopneumatic microvalve: > 150 ms, ref. [38], *** bimetal microvalve: 100–300 ms, ref. [36]

Actuation principle	Type	Size (mm^2)	Max. flow rate / max. pressure difference (N_2 gas)	Switching time (ms)	Power / voltage (DC)	Ref.
Electromagnetic	Macro design	19 x 19	10,000 Sccm / 1,000 kPa	30	< 2 W / -	[76]
Electromagnetic	Si, n.o.	6 x 8	20 Sccm / 10 kPa	0.4 *	< 1 W / < 20 V	[68]
Inverse piezoeffect	Si, n.o.	10 x 10	2,500 Sccm / 600 kPa	2–5	- < 200 V	[56]
Electrostatic	Si, n.o.	6 x 6	400Sccm / 600 kPa	1	< 18 mW/ < 170 V	[65]
Thermopneumatic	Si, n.o. and n.c.	6.3 x 6.6	10,000 Sccm / 700 kPa	>150 **	< 1.5 W -	[42]
Bimetal	Si, n.c	10 x 10	500 Sccm / 600 kPa	100 – 300 ***	< 1.5 W / < 24 V	[40]

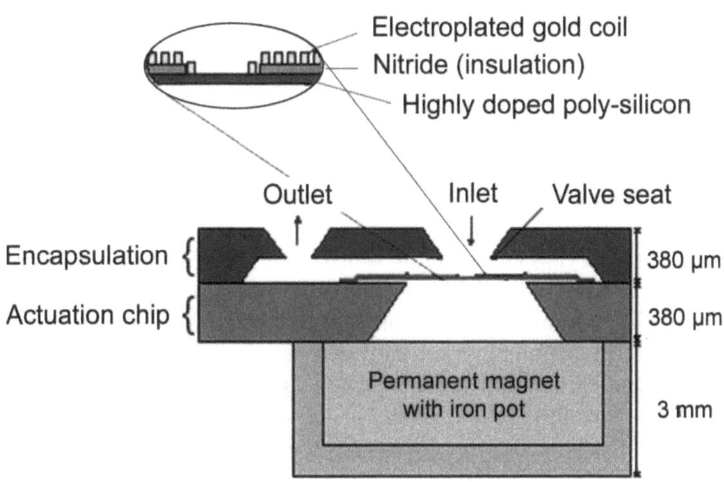

Fig. 2.3. Schematic cross-section of an electromagnetic microvalve in silicon technology [68]

2.2.2 Piezoelectric Microvalves

Piezoelectric actuators may be integrated in microvalves either in a hybrid manner by a bonding step or monolithically by various deposition processes (e.g., sol-gel process, magnetron sputtering, MOCVD) [78–80]. While piezoelectric bending elements allow relatively large deflections, the pressures achieved are relatively small. The latest valve developments on the basis of piezo disks show promising specifications, see Table 2.3. A schematic diagram of a 2/2 normally open microvalve is shown in Fig. 2.4 [56]. In this case, the valve seat and piezo disk are designed with quadratic geometry.

Drawbacks include the hysteresis behavior of the deflection characteristics, the high voltage consumption, and the aging behavior of piezoelectric actuators, which is reflected by, e.g., a gradual decrease in the coupling coefficients over longer terms. The deficit of high operating voltages can be improved largely through the use of thin piezoelectric layers, e.g., by piezo stacks. An advantage of piezoelectrically operated microvalves are short switching times, which are in the range of a few milliseconds.

2.2.3 Electrostatic Microvalves

So far, electrostatically driven microvalves have mainly been produced monolithically using silicon technology. Designs range from electrostatically driven bending elements [57] to bridge structures with pressure compensation [60, 65] and membrane structures [58, 61, 64]. The major drawback of electrostatic drives is the quadratic dependence of force on the capacitor gap, which only permits relatively small deflections. By means of bending elements and membranes, maximum deflections may be increased at the expense of the maximum controllable pressure. Drawbacks include comparably high voltages, charging effects, and nonlinear characteristics which aggravate proportional control. Still, electrostatic microvalves have a relatively simple setup, which can be well manufactured and miniaturized.

Recent valve developments show remarkable specifications, see Table 2.3. A schematic cross-section of an electrostatic microvalve is shown in Fig. 2.5 [65]. It consists of two micromachined silicon chips that exhibit a distance of 5 μm in the actuator area. The plate chip has a movable part in its center that is deflected toward the second chip. An isolation layer of 2 μm-thick SiO_2 prevents a short circuit when the two chips are in contact.

2.2.4 Thermopneumatic Microvalves

Thermopneumatic microvalves have been manufactured both monolithically using silicon technology [38, 41, 42], as well as with the LIGA process by hybrid integration [43]. The central component is a pressure chamber with a movable membrane, which is deflected by thermal expansion due to resistance heating or optical

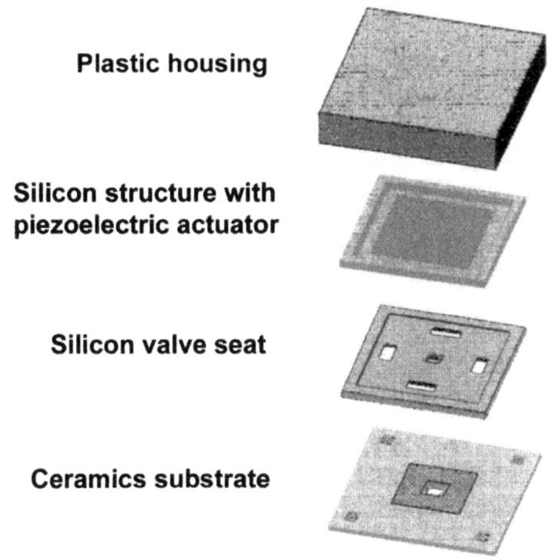

Fig. 2.4. Schematic diagram of a piezoelectric microvalve in silicon technology [56]

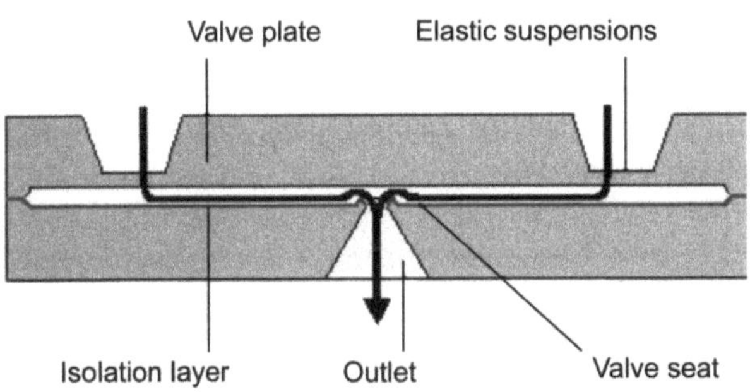

Fig. 2.5. Schematic cross-section of an electrostatic microvalve in silicon technology [65]

heating. Extremely large deflections and forces are achieved by filling the actuator chamber with an actuation fluid with a low boiling point [38, 41, 42]. The opening temperature is set by selecting an appropriate boiling point of the actuation fluid. However, manufacture is aggravated by the tight sealing required. Characteristic drawbacks include relatively high heating powers and long switching times. During heating, these switching times mainly depend on the thermal capacity and the heating power. During cooling, they are dependent on the heat transfer to the environment. Still, the maximum controllable pressures and flows can be set in a wide range. This valve concept has already lead to a commercially available product [81]. Applications cover certain gases as well as various fluids.

Fig. 2.6 shows a schematic of a normally-open thermopneumatically actuated microvalve [42]. The key component is a silicon diaphragm made by etching a recess in a silicon substrate. A Pyrex wafer with a resistor fabricated onto its surface covers the liquid-filled recess, creating a hermetically sealed control cavity. Dissipating energy in the resistor heats the liquid, increasing the pressure inside of the sealed vessel and thereby increasing the volume of the vessel. Thus, the silicon membrane is pushed toward the valve seat. Typical specifications of the valve filled with an actuation fluid are presented in Table 2.3.

To reduce switching times and minimize energy consumption, thermopneumatic membrane microvalves with a buckling mechanism have been developed using both silicon technology [46, 49, 51] and plastic molding [47, 48, 50]. The buckling mechanism is implemented by applying a pre-stressed membrane, whose stress state is set by making use of the different thermal expansion coefficients of membrane and substrate. Buckling is triggered by heating the pressure chamber above a critical value. In this way, two stable switching states can be maintained without supplying energy.

2.2.5 Bimetal Microvalves

Microvalves manufactured monolithically using silicon technology and equipped with a bimetal drive are also commercially available [82]. In most cases, the bimetal drive consists of a silicon membrane and an aluminum structure. As for all thermal principles, relatively high heating powers are required for operation. The switching times achieved may be relatively long depending on thermal capacity and heat-transfer characteristics.

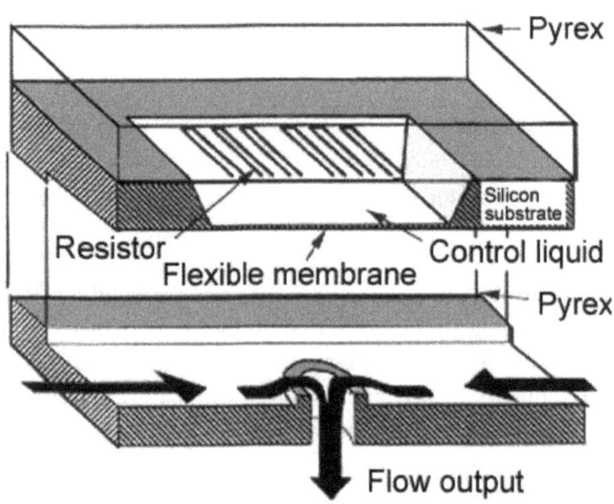

Fig. 2.6. Schematic of a thermopneumatic microvalve in silicon technology [42]

A schematic cross-section of a bimetallically actuated microvalve is shown in Fig. 2.7 [40]. In this case, the valve is designed as a three-way valve with three ports. The valve has two actuation states. In the off state, port 1 is sealed by the pre-stressed actuator membrane and port 2 is connected to port 3. By turning into the on state, port 1 is released and port 3 is sealed. By this design, relatively high controllable pressures at moderate flow rates are reached, see Table 2.3.

Meanwhile, bimetal microvalves have been developed for operation with both gases [33, 34, 36, 38, 40] and liquids [35, 37, 39].

2.3 Linear Actuators

Linear actuators generate linear movements and forces in a given direction and, thus, differ from rotation, torsion, or bending actuators. Linear displacements within or perpendicular to the plane of substrate may be utilized for, e.g., switching, periodic movements or positioning functions. While nonlinear force-displacement characteristics are favorable for switching, positioning functions require flat characteristics with a small variation of force. By combining linear actuators with other mechanical microcomponents, displacement sensors, and control units, mechanical microsystems are obtained, which, e.g., enable a precise handling of small objects or specifically influence optical beam paths. Potential fields of application are in industrial automation technology, telecommunication, and computer technology.

Linear actuators that have been developed for microsystems applications so far are based on magnetic [25, 83–93], electrostatic [94–115], piezoelectric [116–122], magnetostrictive [123, 124], and thermal principles [125], for instance. Fluidic linear actuators [27, 126] will not be considered, as they generally require an external pressure supply. Table 2.4 presents the major specifications of selected linear actuators reflecting the current state of the art.

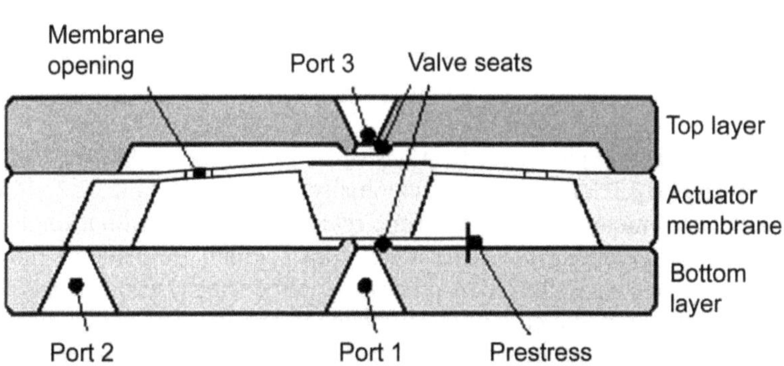

Fig. 2.7. Schematic cross-section of a bimetal microvalve in silicon technology [40]

Table 2.4. Specifications of linear actuators fabricated by LIGA technology. N denotes the number of windings of the coils. The maximum force is limited by the adhesion of the LIGA structures onto the substrate and, therefore, is expected to be far below the total force available of about 1 N

Actuation principle	Type	Size (mm^2)	Max. force / Max. displacement	Switching time (ms)	Power / voltage (DC)	Ref.
Reluctance	LIGA, a single integrated coil, N = 40	4 x 4	3 mN / 190 μm	3	220 mW / 1.3 V	[87]
Reluctance	LIGA, two hybrid coils N = 500	5 x 5	1 mN / 450 μm	1	2 x 80 mW / 2 V	[85]
Electrostatic	LIGA, conic capacitor elements	4.2 x 4.2	10 mN / 80 μm	4	< 1 mW / 70 V	[111]
Inverse piezoeffect	LIGA on piezo substrate	10 x 1	-* 140 μm	0.5	< 1 mW / 1,000 V	[120]

* No data available

2.3.1 Magnetic Linear Actuators

Nearly all magnetic linear actuators fabricated by microengineering so far are based on the reluctance principle. Such actuators mainly consist of three components: a fixed magnetic core for conducting the magnetic flux, a movable anchor, and a spring for guiding and resetting the anchor. Up to now, electrodynamic principles have met only with little interest, as the required three-dimensional coil geometries are difficult to implement. Of particular interest are actuators with microstructured permanent magnets, due to the more favorable scaling behavior of force and work density [83, 127]. Thus, the fabrication of hard magnetic microstructures is investigated extensively in various research institutions [127, 128]. To reach high forces and large displacements in the plane of substrate, high aspect ratios are required, which are achieved, for instance, by magnetic LIGA actuators [25, 84–91].

Table 2.4 presents typical specifications of two magnetic reluctance actuators with an integrated LIGA coil or two externally wound coils. Fig. 2.8 shows a scanning electron micrograph of a magnetic reluctance actuator with a single integrated coil [87]. Integrated microcoils with an axis within the plane of substrate have been fabricated using the LIGA process by introducing a second LIGA layer. The coil is wound around a core of electroplated permalloy. The core is firmly attached to the substrate and guides the magnetic flux to the gap between core and keeper. The latter is suspended by springs and is, therefore, movable. The core's height is 150 μm. All microstructures have been made through batch fabrication.

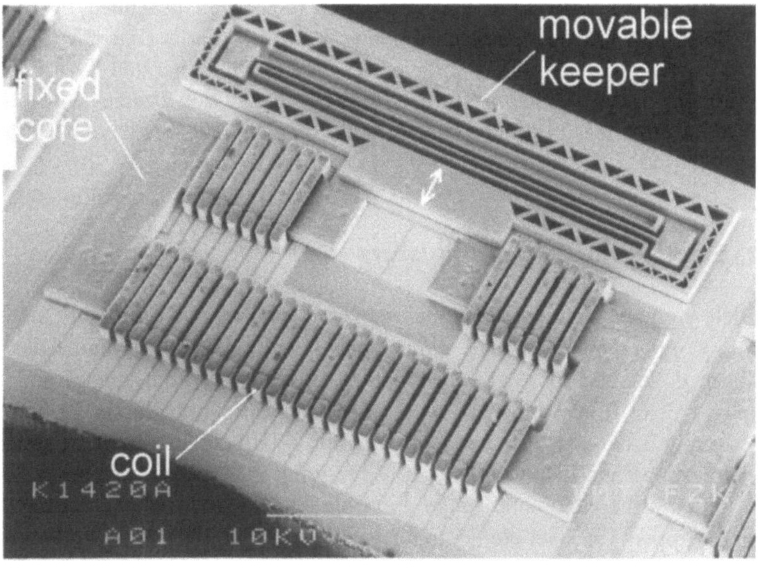

Fig. 2.8. Scanning electron micrograph of a magnetic linear actuator in LIGA technology [87]

The maximum force is about 3 mN for a current of 100 mA. Due to the process-inherent high coil resistance, the control power is 220 mW, in spite of the small number of windings.

In principle, reduction of maximum displacement allows the increase of the maximum force encountered at the end of the path at minimum air gap. Such linear actuators are of interest to microrelay applications needing contact forces of at least 10 mN [86, 89]. However, a constant current must be applied to maintain the contact. Control voltages are below 3 V and, hence, are compatible with microelectronic circuits. Switching times are in the range of milliseconds, which is of interest to both switching and positioning functions. Magnetic reluctance actuators may be used simultaneously as inductive position sensors [85, 86]. Thus, interesting options of implementing compact positioning systems are obtained.

Technical problems arise above all when establishing three-dimensional microcoils to generate magnetic fluxes and fabricating integrated magnets of high saturation field strength and high permeability. The number of coil windings with an axis within the plane of substrate is strongly limited by process-inherent boundary conditions. The fabrication of such microcoils is associated with numerous drawbacks, such that hybrid integration of externally wound microcoils is favored from today's point of view [84, 85, 89, 90, 93]. Using electrodeposition, soft magnetic iron-nickel alloys with saturation flux densities of 1 Tesla and a permeability in excess of 1,000 can be produced [129]. In general, the development of magnetic materials with a small magnetostriction coefficient is desirable to minimize internal stress.

Recently, magnetostrictive linear actuators have been produced by thin-film technology [123, 124]. The actuators consist of a foil that is coated with magnetostrictive stripes on both sides. Linear movements are generated by a travelling wave. Small pins on the lower surface improve the friction contact with the base plate. So far, both feasibility and functionality of such actuators have been demonstrated.

2.3.2 Electrostatic Linear Actuators

In the late eighties, electrostatic actuators were proposed as a promising method of generating movements in small dimensions [130]. Since then, numerous types have been developed for a variety of applications using both silicon and LIGA technology. The first prototypes were produced from polysilicon [94]. Due to small structural heights of a few micrometers, relatively small forces in the micronewton range and displacements of a few micrometers are achievable. Spatial resolution is in the nanometer range. Possible applications include, e.g., actuators for microgrippers [96], set elements for vertical orientation of optical components [113], and positioning elements for fine positioning of tips in scanning force or scanning tunneling microscopes [98, 102]. Apart from horizontal actuators, linear actuators with movement direction perpendicular to the plane of substrate have been developed in silicon technology [108]. By means of the LIGA process, electrostatic linear actuators with large structural heights and aspect ratios have been developed. These actuators are characterized by high forces and displacements within [103, 104, 106, 111] and perpendicular to [107] the substrate layer. As a possible field of use, optical microswitches were studied [112]. Using a special type of LIGA process for sub-micrometer dimensions, electrostatic linear actuators with minimum structural dimensions of 500 nm and an aspect ratio of 60 have been fabricated [109, 110].

Alternative mechanisms generating linear motions are electrostatic inch-worm drives and so-called scratch-drive actuators, for instance. Inch-worm drives generate step-like linear motions by a sequence of clamping and translation movements coordinated in time [114, 115]. Scratch-drive actuators have been developed above all for self-assembly of optical components in optical microsystems. The underlying mechanism is based on a periodic sequence of electrostatically triggered bending and shifting movements [105].

Table 2.4 lists typical specifications for an electrostatic linear actuator fabricated through LIGA technology consisting of a two-dimensional array of conic capacitor elements. A scanning electron micrograph of this actuator is shown in Fig. 2.9. The linear actuators mainly consist of three components: an array of fixed capacitor elements, the corresponding movable capacitor elements, and springs for suspension and resetting the movable elements. The movable capacitor elements are interconnected by a frame to provide sufficient rigidity. Large forces and displacements are achieved by designing a large number of capacitor elements with high aspect ratio.

Fig. 2.9. Scanning electron micrograph of an electrostatic linear actuator in LIGA technology

The conic design of the capacitor elements allows particularly small capacitor gaps during operation. In the present case, the height of the capacitor elements is 120 µm. A maximum displacement of 80 µm and a maximum force of 10 mN are reached at 70 V operating voltage and a minimum capacitor gap of 2 µm. In principle, the displacement can be further increased by allowing a lower maximum force. Thus, displacements of 190 µm and maximum forces of 1 mN can be reached at the same structural height and gap width [103]. In general, spring elements of sufficient transverse stiffness are required to take up the electrostatic transverse forces occurring perpendicular to the movement direction during operation. Electrostatic transverse forces result from electrostatic forces of plate attraction in the capacitor array, which coexist with the forces of plate alignment. In the case of small capacitor gaps, the forces of plate attraction dominate and, hence, might cause a short circuit. To prevent this, the actuator design must be provided with stops or other countermeasures. Switching time of the electrostatic linear actuator is in the millisecond range and, hence, of interest to both switching and positioning functions. Electrostatic linear actuators may be used simultaneously as capacitive position sensors, see, e.g., [131]. In this way, compact positioning systems can be realized.

Due to their relatively simple geometry, electrostatic linear actuators can be produced using standard planar technologies. The high input resistances, however, result in relatively high voltages required for operation. This may give rise to compatibility problems with microelectronic circuits. Therefore, an attempt was made to minimize the voltage consumption of electrostatic linear actuators at the expense of forces and displacements, which is, however, only possible for special applications [96, 100]. Charge accumulation on insulating components may cause failure or complete destruction [132]. Other characteristic drawbacks result from the high sensitivity to various ambient impact, such as humidity, dust particles, electromagnetic fields, etc. This is why electrostatic linear actuators have to be provided with appropriate housings. The effort in terms of assembly and bonding technology increases accordingly. Contrary to silicon actuators, electroplated linear actuators made of Ni exhibit comparably bad mechanical properties. This is reflected in the low plastic deformation limit of reset springs or minor quality factors during dynamic operation [111].

Comparison of electrostatic and magnetic linear actuators in terms of work density and switching time yields similar results. In macroscopic design, magnetic linear actuators reach a far higher work density than electrostatic linear actuators. This is confirmed by the scaling behavior of both actuation principles given in Table 2.2. It is explained in Sect. 2.1 that the unfavorable scaling behavior of magnetic actuators leads to a stronger decrease of the achievable work density in the case of miniaturization. Therefore, magnetic actuation principles become uninteresting in the sub-millimeter range.

2.3.3 Piezoelectric Linear Actuators

Piezoelectric linear actuators in miniature design have already been applied in a number of fine positioning systems, as they are characterized by a high displacement resolution, high forces, and short reaction times [118, 133]. These actuators are integrated in a hybrid manner using appropriate assembly and bonding technologies. The latest activities increasingly focus on the monolithic fabrication of piezoelectric linear drives by further developed deposition processes, e.g., sol-gel processes [117]. Microstructuring piezoelectric layers allows the fabrication of driving and gear elements from a single workpiece, which results in new options for actuator development [116]. The maximum achievable displacements of piezoelectric linear actuators are relatively small, which meanwhile has lead to the development of various mechanisms for increased displacement, see, e.g., [116, 120]. Unlimited displacements are reached by piezoelectric ultrasonic motors, where propagating surface waves are transmitted by friction forces [121] or by piezoelectric inch-worm drives [122].

Fig. 2.10 shows a piezoelectric linear actuator fabricated by electrodeposition of LIGA bending elements on a piezo substrate [120]. The main specifications of this actuator are summarized in Table 2.4. In this case, the displacement is increased by a LIGA spring with a preferred direction of deflection produced by a sacrificial layer technique. This allows an increase of displacement by a factor of 30. Maximum displacements amount to 140 μm. The achievable forces are limited by the adhesion of the LIGA structures onto the substrate base. Therefore, appropriate adhesion surfaces of sufficient dimensions are required.

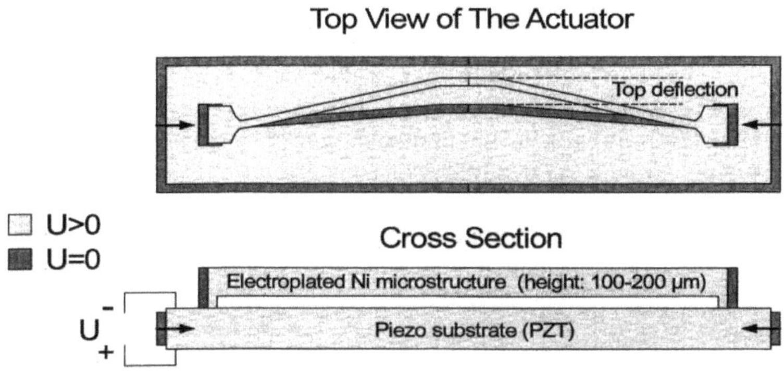

Fig. 2.10. Schematic diagram of a piezoelectric linear actuator consisting of a LIGA beam on top of a piezoelectric substrate [120]

As far as dynamics and power consumption are concerned, the piezoelectric linear actuator comes off best. However, due to the large layer thickness of the piezo substrate of 500 μm, very high electric voltages of up to 1,000 V are required for operation. Voltage consumption can be reduced by using a multilayer substrate. However, this results in a considerable additional effort required for substrate production. Another drawback of using a multilayer substrate is the higher electric capacity, which is associated with increased charging currents during dynamic operation and may lead to heating effects.

Piezoelectric linear actuators exhibit hysteresis and creep effects, which may lead to positioning errors of at least 3%, in spite of material optimization [119]. Higher accuracies require appropriate methods such as control with sensors in closed loops. Other problems result from the pressure and temperature dependence of the piezoeffect, which cause undesired cross sensitivities, e.g., in the presence of stress or by heating adjacent components. High field strengths or aging may lead to depolarization and, hence, functional failure.

2.4 Development Potential of SMA Microactuators

Table 2.5 summarizes the major properties of SMA microactuators. The work density of SMA actuators exceeds that of other actuation principles by at least an order of magnitude, see Table 2.1. For microsystems applications, the properties of "direct actuation with an electrical current" and "simple design" are of particular importance. To operate SMA actuators, only small electric voltages are required that are compatible with microelectronics. When taking material-dependent load limits into account, fatigue effects can be avoided even at high cycle numbers and a high reliability can be achieved. As SMA actuators do not only fulfill transducer functions, but also various other functions such as electrical, thermal or structural functions, compact designs are possible, which allow easy fabrication and a large degree of miniaturization.

When comparing various actuation principles, generally the long response time of SMA actuators is noted. As in all thermal actuators, it results from the low heat transfer rates. Other possible drawbacks concern high power consumption and small efficiency, since during the operation of SMA actuators only a maximum fraction of a few percent of supplied thermal energy is converted into work. However, it must be emphasized that these drawbacks lose significance with increasing miniaturization.

As obvious from Table 2.2, work density of SMA actuators remains constant upon miniaturization, while it decreases quadratically for reluctance actuators, for instance. In the most favorable case of maximum allowable current density scaling according to $J \propto r^{-1}$ [26], the power density of SMA actuators even increases quadratically. In this case, heating and cooling times decrease quadratically, such that a considerable improvement of dynamic actuation performance in small dimensions has to be expected. The efficiency of SMA actuators will not change significantly upon miniaturization. However, absolute power consumption drops to the milliwatt range, such that power requirements become acceptable for most applications.

It can be concluded from the scaling behavior of various actuation principles that the advantages of SMA actuators will gain significance upon miniaturization, while the drawbacks will become less important. In this respect, SMA actuators are virtually destined for applications in microsystems. Successful development of ink jet printers convincingly demonstrates that thermal actuation principles in general are highly attractive for microsystems applications.

Table 2.5. Properties of SMA microactuators

Advantages:	High energy density Electrical actuation at small voltages Simple designs High reliability Variety of shape changes Noise-free operation Biocompatibility
Drawbacks:	Low efficiency High power consumption Long response time Temperature range is alloy-dependent Fatigue effects

3 Shape Memory Effects

The common aim is in the frame of a "biomimetic" inspiration to engender materials and structures with the ability to respond adaptively to their environmental constraints and their own condition.

–P.F. Gobin and B. Culshaw,
in *Preface of 3rd International Conference on Intelligent Materials*

Understanding shape memory requires a thermodynamic approach that takes into account the structure of the solid and the mechanical properties resulting from it. For this reason, the crystallographic and thermodynamic aspects of shape memory effects will be discussed below to the extent to which they are relevant to the development of microactuators. Detailed descriptions of this subject have been published in numerous review articles and books such as [134–137]. Particular emphasis will be placed on the special properties of NiTi-based alloys.

3.1 Overview

The term "shape memory" describes the unusual ability to remember shape, which can be initiated in certain materials either thermally or mechanically. Even after heavy deformation, materials with shape memory are able to recover a previously memorized shape. This phenomenon was discovered in brass alloys as early as the late thirties. However, the real importance of the effect has become obvious only since its discovery in a NiTi alloy [138], where it is particularly pronounced.

In the meantime, a number of materials systems with shape memory properties are known [139], see Table 3.1.

Table 3.1. Material systems with shape memory properties

Metals:	NiTi-based alloys: NiTi, NiTiCu, NiTiPd, NiTiFe, ... Cu-based alloys: CuZn, CuZnAl, CuAlNiMn, ... Fe-based alloys: FePt, FeMnSi, FeNiC, ...
Polymers:	e.g., PTFE (polytetrafluoroethylene)
Ceramics:	e.g., ZrO_2
Biological systems:	e.g., bacteriophages

The most important materials of commercial significance can be classified either as metal alloys or as polymers. Furthermore, there are ceramics and biological systems in which shape memory properties are observed as well. An example to be mentioned in this respect are bacteriophages, which use a shape memory mechanism when entering host cells. SMAs are currently the focus of interest, as they have proved to function in a number of applications and show an unforeseeable potential for future applications. Therefore, this book will be limited to SMAs.

In contrast to conventional structural materials, in which shape changes are made up of elastic, plastic or thermal contributions, SMAs show three additional types of shape changes that are associated with shape memory characteristics. These effects are illustrated in Fig. 3.1 by the example of a helical spring:

(a) One-way effect: After removal of the load $\vec{F}$, the helical spring shows permanent deformation. This seemingly plastic deformation recedes completely upon heating.
(b) Two-way effect: In addition to the one-way effect, there is also a defined shape change upon cooling.
(c) Pseudoelasticity: Mechanical loading, $\vec{F}$, expands the helical spring to a large extent. When the load is removed, the spring still returns to its initial shape.

In most cases, the underlying mechanism is based on a reversible, martensitic transformation between a high-temperature austenitic phase and a low-temperature martensitic phase well below the melting temperature. The atoms in this case perform coordinated movements, dominated by shear, which can lead to macroscopic shape changes. Other transformations associated with shape memory, for instance, are rhombohedral (R-) phase transformations [140] and, subject to some limitations, bainitic transformations [141]. This overview will be restricted to martensitic transformations.

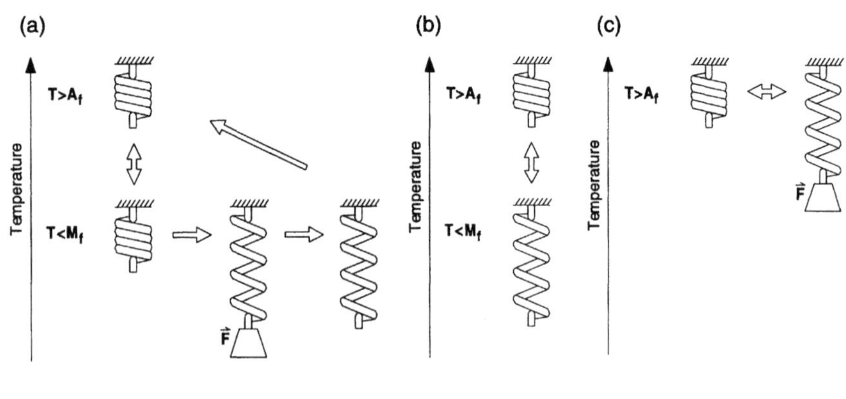

Fig. 3.1. Shape memory effects shown by the example of an SMA helical spring; (a) one-way effect, (b) two-way effect, (c) pseudoelasticity

Martensitic transformations are associated with hysteresis. As a consequence, four temperatures are generally required for characterization. Transformation to austenite begins at the austenite start temperature, A_s, and ends at the austenite finish temperature, A_f. Accordingly, transformation into martensite begins at the martensite start temperature, M_s, and finishes at the temperature M_f.

Reversibility in the crystallographic sense implies that the austenite-martensite transformation proceeds in both directions in the opposite sequence of steps, without giving rise to any defects. Thus, exactly the same lattice structure is restored at the end of a transformation cycle. This case is also referred to as ideal thermoelasticity. Irreversible processes such as diffusion, formation of dislocations, or slipping of lattice planes do not occur in this case. The shape memory effects are usually illustrated in stress-strain-temperature (σ–ε–T) diagrams, see Figs. 3.2–3.4.

3.1.1 One-Way Effect

In the martensitic state (Fig. 3.2, $T<M_f$), the material exhibits a very low elastic limit. Elastic straining is followed by a pseudoplastic strain range in which the component can be strained reversibly by up to several percent. Further strain behavior is characterized by plastic deformation up to fracture. After relaxation in the pseudo-plastic range, an apparent deformation, ε_{1W}, is retained.

By heating the deformed component above the austenite finish temperature, A_f, a complete shape recovery is possible. The maximum reversible strain, for instance of NiTi single crystals in the [233] direction, is 10.7% [142]. As this effect only occurs upon heating, i.e., only in one direction, it is referred to as a one-way effect. Renewed cooling and heating no longer changes the shape. The original memorized shape is imprinted upon the component prior to loading by heat treatment. If the shape recovery during reverse transformation to austenite is hindered, high forces occur that can be used to perform work. This is the basis of SMA actuators.

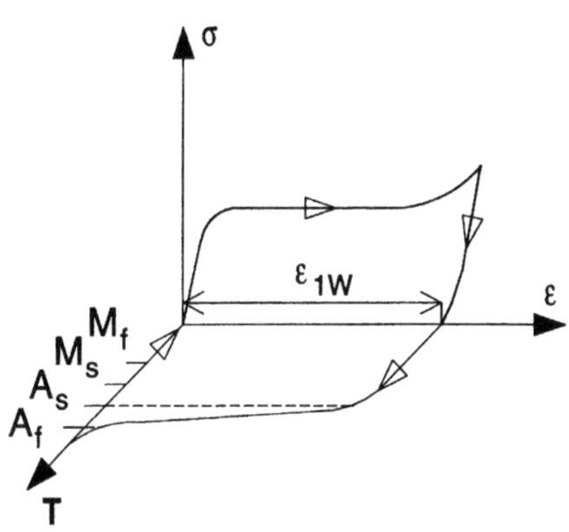

Fig. 3.2. Stress-strain-temperature characteristic for the one-way effect

3.1.2 Two-Way Effect

The two-way effect is associated with a shape change upon heating and cooling without requiring any external load. This gives rise to a strain characteristic located within the ε–T plane, see Fig. 3.3. The shape change can be repeated without renewed deformation. However, in principle, the shape change attained is less pronounced than with the one-way effect. The shape change upon cooling is achieved by imprinting ordered internal stress fields on the material. The underlying mechanism is based on the formation of preferred martensite variants and is described in greater detail in Sect. 3.2.2. One special case of the two-way effect is the all-round effect caused by the formation of Ti_3Ni_4 precipitates with a preferred orientation [143].

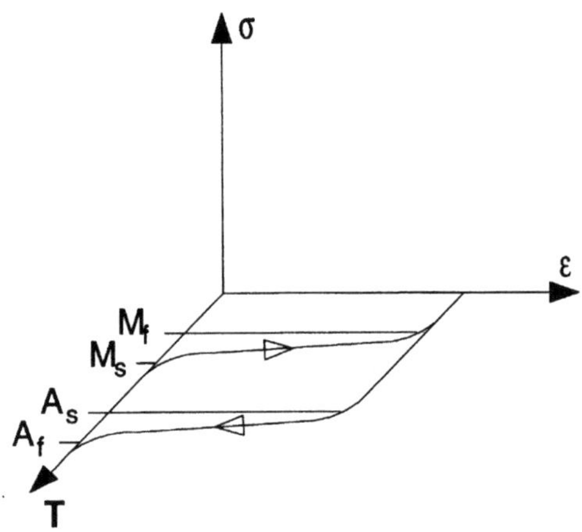

Fig. 3.3. Strain-temperature characteristic for the two-way effect

3.1.3 Pseudoelasticity

In the austenitic state ($T>A_f$), the material exhibits pseudoelastic behavior, see Fig. 3.4. In contrast to previous effects, no temperature change is required in this case. The strain characteristic is, therefore, located in the σ–ε plane. Above the elastic limit, there is a plateau in which a highly nonlinear deformation occurs up to a virtual yield limit, ε_{pe} . Above ε_{pe} , there is plastic deformation up to fracture. If the component is loaded only as far as ε_{pe} , unloading passes through the lower hysteresis loop with the strain disappearing completely. In polycrystalline NiTi, the maximum reversible strain can be 7 to 8% [144], and in some Cu-based SMAs, up to 18% [145]. The plateau region is caused by stress-induced transformation of austenite into martensite, see Sect. 3.2.2.

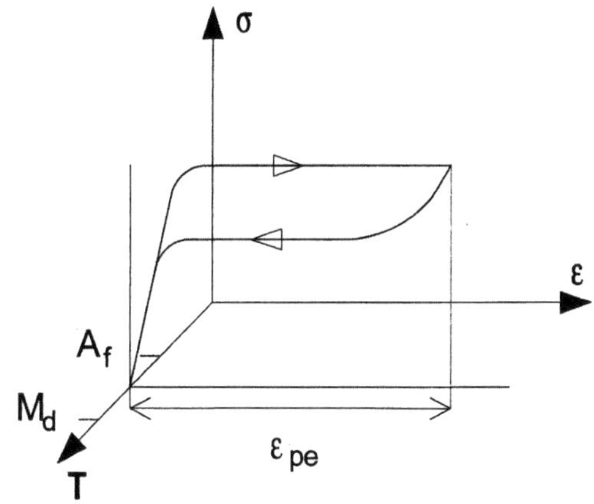

Fig. 3.4. Stress-strain characteristic for pseudoelasticity

For the stress-induced formation of martensite there is an upper temperature limit, M_d, above which competing irreversible processes, such as the formation of dislocations and slipping, are favored thermodynamically. The temperature window, $M_d>T>A_f$, in which pseudoelastic behavior occurs can be set by various thermomechanical processes [146]. Above M_d, SMAs behave like conventional materials with elastic strain characteristics and subsequent plasticity up to fracture. Besides the pseudoelasticity described here, which is due to martensitic transformation, the much less pronounced effect of twin-pseudoelasticity is described in the literature [147].

In general, it is possible that the three shape memory effects occur either individually or superimposed upon each other, depending on the temperature and the stress situation. Below the A_f temperature, for instance, a combination of stress-induced and thermally induced martensitic transformations is present. The conditions under which the effects occur also depend on chemical composition, the ordering parameter of the crystalline phase, and on the structure of the SMA.

3.2 Crystallographic Aspects of Martensitic Transformations

3.2.1 Microscopic Structural Changes

One characteristic feature of martensitic transformations is the coordinated movement of atoms in the crystal lattice. In the high-temperature phase, the crystal lattice has a higher symmetry than in the low-temperature phase. The transition between the two states occurs by coordinated movements of the lattice atoms. Although the relative movements of atoms are small compared to the interatomic distance, it is possible that a macroscopic shape change results.

Figure 3.5, for example, shows the crystal structure of NiTi in austenitic and martensitic conditions. The austenite has a space-centered cubic lattice structure of the B2 type. The structure of martensite corresponds to that of a monoclinically distorted shape of AuCd (B19' type). In austenitic state, Fig. 3.5 (a), a tetragonal cell is shown that is indicative of the relationship with martensite. The lattice parameters of both phases are listed in Table 3.2. Comparison of the dimensions indicates the size of the coordinated shear movement performed by the atoms in the austenite-martensite transformation.

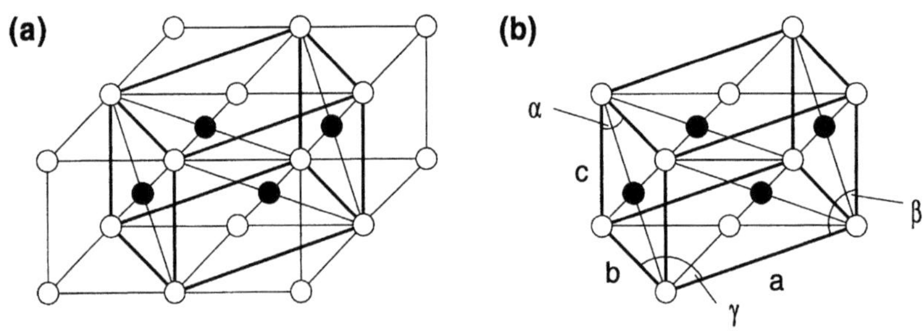

Fig. 3.5. Four adjacent unit cells of the crystal lattice of NiTi in the austenitic state (a) and a monoclinically distorted unit cell in the martensitic state (b)

Table 3.2. Lattice parameters of NiTi [148]

Lattice parameter	Martensite	Austenite
a	0.2898 nm	0.3015 nm
b	0.4108 nm	0.4263 nm
c	0.4646 nm	0.4263 nm
α	90°	90°
β	97.78 °	90°
γ	90°	90°
V	54.79×10^{-3} nm^3	54.78×10^{-3} nm^3

Martensitic transformations can be represented as a sequence of lattice deformation, lattice-invariant shear, and lattice rotation. This sequence is shown in Fig. 3.6. As the martensitic transformation is a first-order process thermodynamically, martensite is produced in the ambient austenite lattice by nucleation and growth. In this process, the austenite lattice is locally deformed by the shear forces. The occurrence of high shear forces, the diffusionless nature of the transformation, and the fact that the transformation occurs inside a parent crystal result in high strain and, consequently, high energy densities counteracting martensite formation.

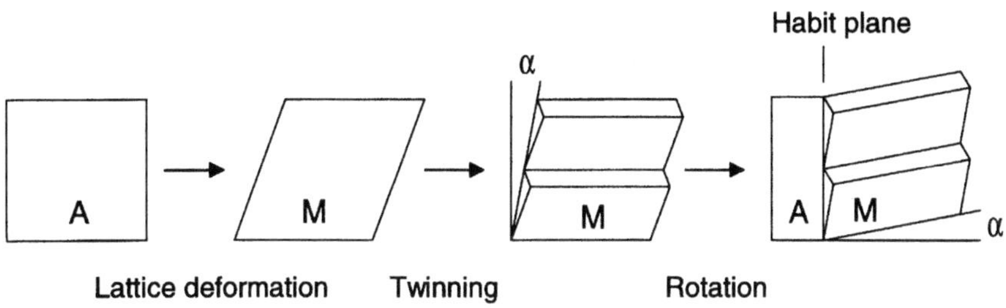

Fig. 3.6. Two-dimensional schematic diagram of the microscopic subprocesses in martensitic transformation

High strain can give rise to various relaxation mechanisms. In principle, there is the possibility of dislocation slipping or twinning. In both cases, the structure of the martensite lattice is preserved (lattice invariance). However, only twinning is reversible. This is why SMAs normally only show the process of lattice-invariant twinning (self-accommodation). In this way, shearing of the twins with different orientations is averaged out. The mechanism of self-accommodation not only causes the formation of lattice-invariant planes between the martensite and the austenite lattices (habit planes), but also gives rise to a clear decrease of straining. Consequently, when cooling without external loads, there is no shape change associated with the one-way effect.

Figure 3.6 shows the production of deformation twins arranged in stacks of long periods. In the two-dimensional case, there are two different martensite variants, which emerge from the austenite lattice by shearing in two possible directions and are interconnected by twin boundaries. In three dimensions, there is a correspondingly larger number of martensite variants. Assigning two possible directions of shear to each of the six fundamental planes gives rise to a total of twelve correspondence variants. In the case of NiTi, this produces a total of 24 possible habit planes. Depending on the position of the lattice vectors relative to the shear plane and the direction of shearing, two different types of symmetry of deformation twins are distinguished [149]. In most cases, rotation is necessary in addition to twinning in order to ensure continuity of the habit planes.

The macroscopic result of the sometimes complicated microscopic shape change associated with martensitic transformations is lattice-invariant strain relative to the habit plane, which can be described by shearing and volume dilation, $\Delta V/V$ [150]. This is illustrated in Fig. 3.7, where $m\vec{d}$ stands for the vector of lattice-invariant strain.

As the volume change in SMAs is very small, typically $\Delta V/V = 0.3\%$, lattice-invariant strain nearly corresponds to simple shearing. Lattice-invariant strain results in the maximum reversible strain in the shape memory effect. Consequently, this limit can be predicted theoretically for single crystals on the basis of the orientation of the high-temperature phase. Figure 3.8, for example, shows maximum reversible strains of martensite in NiTi as a function of crystal orientation.

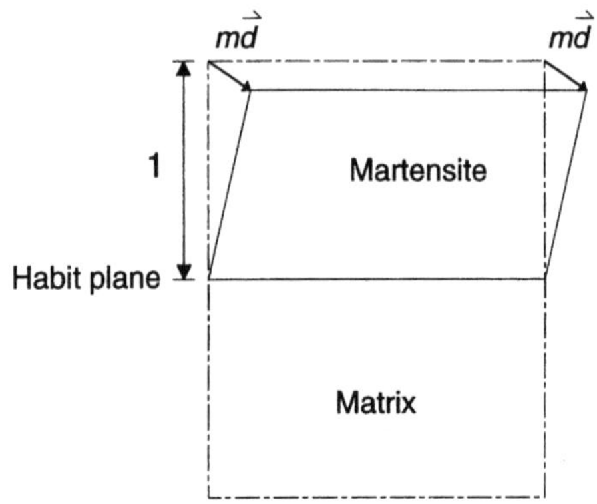

Fig. 3.7. Two-dimensional schematic diagram of lattice-invariant strain

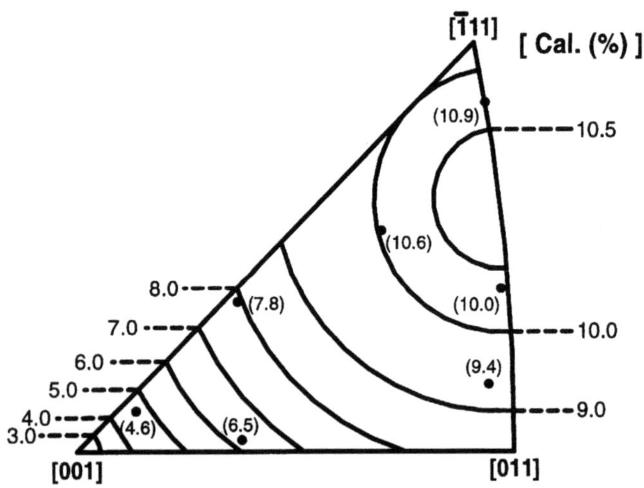

Fig. 3.8. Orientation dependence of the maximum reversible strain for the martensitic transformation in NiTi. The contour lines were calculated, the figures in brackets are experimental data [151]

3.2.2 Structural Changes in the Presence of Stress

An external stress field can easily move twin boundaries. Consequently, martensite variants with an energetically favorable orientation relative to a given stress field are preferred and can grow at the expense of the adversely oriented variants (mechanism of variant coalescence) [152]. The generally accepted rule of variant selection indicates that the variant is favored which results in the highest transformation strain at a given external stress [152]. In the extreme case, the entire crystal may consist of only one martensite variant. The maximum reversible shape change occurring in this case represents the maximum possible shape memory effect for the associated crystallographic orientation.

The mechanisms of self-accommodation and variant coalescence involved in martensitic transformation are illustrated in Fig. 3.9. In thermally induced martensitic transformation without any external load, self-accommodation gives rise to combinations of martensite variants, which keep the shape unchanged. Subsequent external loading in martensitic condition ($T<M_f$) causes martensite variants with preferred orientation to grow as a result of variant coalescence. This gives rise to the pseudoplastic development of the stress-strain characteristic shown in Fig. 3.2, which continues until only martensite variants with preferred orientation are left. Depending on the type of external loading, it is possible in this way to produce numerous macroscopic shape variants that remain thermodynamically stable in the martensitic state upon load relief. The more the variety of martensite variants is constrained by external loading the more pronounced the associated macroscopic shape change.

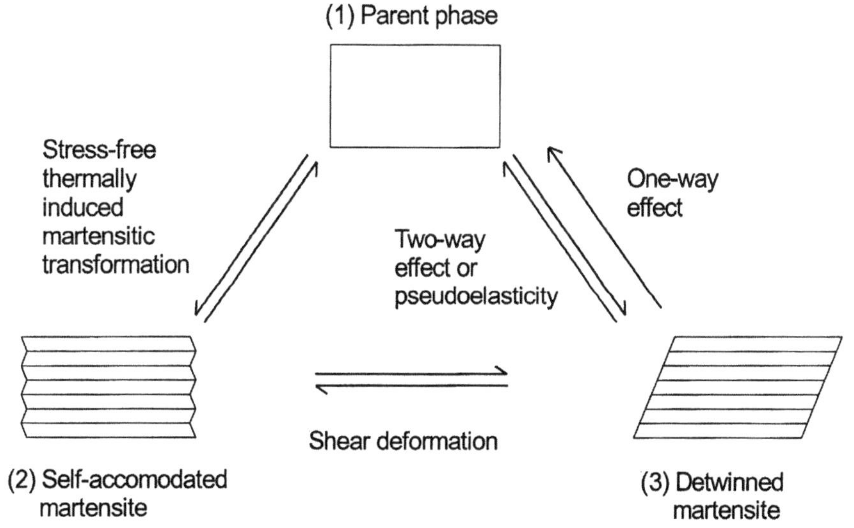

Fig. 3.9. Two-dimensional schematic of the microscopic processes in martensitic transformation:

One-way effect: $(1) \xrightarrow{\text{cooling}} (2) \xrightarrow{\text{deformation}} (3) \xrightarrow{\text{heating}} (1)$

Two-way effect: $(1) \xrightarrow{\text{cooling}} (3) \xrightarrow{\text{heating}} (1)$

Pseudoelasticity: $(1) \xrightarrow{\text{loading}} (3) \xrightarrow{\text{relaxation}} (1)$

Imprinting stress fields in the material by means of special thermomechanical training techniques also allows the range of martensite variants to be constrained without any external loading. One process frequently applied to NiTi alloys is the generation of directed lattice defects by pronounced deformation in martensitic condition or by thermal cycling under the mechanic load (training). The different training methods have been described in detail in the literature [153]. In trained SMA materials, thermally induced martensitic transformation also results in the formation of a preferred macroscopic shape upon cooling, which is observed as a two-way effect.

The martensitic phase is thermodynamically stabilized by stress well beyond the A_f temperature limit. In the range of $T>M_s$, martensite can also be produced especially by external stress (stress-induced martensite, SIM). The SIM mechanism causes pseudoelastic behavior, see Fig. 3.4. In this case, only certain martensite variants with favorable orientation are already produced during phase transformation. This process continues until the growth of martensite variants of favorable orientation has been completed. Limitation of the variety of martensite variants is again associated with a macroscopic shape change being the more pronounced the fewer martensite variants remain.

3.2.3 Shape Memory

In SMAs, shape memory occurs through reverse transformation from martensite to austenite initiated either by heating or, in the case of pseudoelasticity, by removal of the external stress. As is shown in Fig. 3.9, the original austenitic crystal structure is restored irrespective of the combination of martensitic variants produced, as the high symmetry of austenitic crystal structure admits only one energetically preferred crystal structure. The associated macroscopic shape change is observed as shape memory. In the special case of the two-way effect, the martensite carries an energetically preferred combination of martensitic variants that allows shape memory to occur through the transformation from austenite to martensite as well.

3.2.4 Preconditions of Shape Memory

The discussion above implies various requirements that must be met in SMAs for shape memory to occur:

(1) Compatibility: In the transformation process, the new phase must be able to form within the parent crystal without losing coherence of the crystal lattice. There must be no phase separation due to microfracture. Consequently, the crystal lattices must be similar in terms of atomic distance, symmetry, and coordination number (compatibility). Major changes in volume in transformation have a negative impact.

(2) Structural transformations with the largest possible amount of shear lead to a pronounced memory effect.

(3) Negligible diffusion: The transformation temperatures must be in a range in which the jumping time is sufficiently long for atoms to interchange sites.
(4) Reversibility: The mechanism of lattice-invariant deformation is of decisive importance to the reversibility of martensitic transformation. Crystallographic reversibility is ensured when lattice-invariant deformation occurs by twinning, while the formation of dislocations and stacking faults is avoided. The shape memory effect is more stable the higher the critical stress for the formation of dislocations and stacking faults. Consequently, hardening techniques have been developed in order to raise this stress limit. In polycrystalline NiTi it was possible, in this way, to increase the maximum reversible strain to 7–8% [154].
(5) Control of the transformation pathway: A mechanism is required to translate the transformation strain into a macroscopic shape change. This can be achieved by external or internal stresses, which minimize self-accommodation during transformation and limit the martensite variants, thus maximizing the macroscopic shape change.

3.2.5 Phenomenological Theory of Martensitic Transformations

From the point of view of crystallography, the purpose of a theory of martensitic transformation is to predict all crystallographic parameters associated with transformation. This refers, for instance, to the magnitude and direction of shape change or the position of the habit planes. At the present time, there are two phenomenological theories that allow such predictions to be made. They were developed independently by Wechsler-Lieberman-Read (WLR) [155] and Bowles-Mackenzie [156], but are equivalent [157].

A key point in the WLR theory is the minimization of the strain energy in transformation. As strains are concentrated at the interface between austenite and martensite, the interface is assumed to remain an invariant during transformation, at which the lattice deformations on average cancel out. Consequently, in the WLR theory, the shape strain matrix of martensitic transformation, P_1, is resolved under the condition of the habit plane being a lattice-invariant plane [155]. The shape strain matrix, P_1, is composed of the operators of the lattice deformation, B, lattice-invariant shear, P_2, and rotation, R (see Fig. 3.6):

$$P_1 = RP_2B \tag{3.1}$$

It now follows from the lattice invariance of the habit plane that the shape strain can be described by shearing parallel to the habit plane and volume dilation, $\Delta V/V$, normal to the habit plane,

$$P_1 = I + m_1 d_1 p_1 , \tag{3.2}$$

where I is the (3x3) unit matrix, m_1 describes the amount of shape strain, and d_1 and p_1 describe the unit vectors in the direction of shape strain and normal to the habit plane, respectively. This result corresponds to the plot of lattice-invariant strain shown in Fig. 3.7. The input parameters necessary to determine the crystallographic quantities, p_1, m_1, and d_1, thus are limited to the empirically determined lattice parameters of austenite and martensite, the lattice correspondence between austenite and martensite, and lattice-invariant shear.

3.3 Thermodynamic Aspects of Martensitic Transformations

3.3.1 Gibbs Potentials

The driving force of martensitic transformations is determined by the thermodynamic potentials in austenitic and martensitic condition. Since martensitic transformations progress without diffusion, the chemical composition does not change. Thus, SMAs can be treated as single component systems. Furthermore, time-dependent effects can be neglected in thermodynamic models. By selecting temperature and stress as independent variables corresponding to usual experimental conditions, the thermodynamic potential becomes the Gibbs potential, G, which can be written for both phases as follows:

$$G^A = U^A - S^A T - \{\varepsilon\}^A \{\sigma\}$$

$$G^M = U^M - S^M T - \{\varepsilon\}^M \{\sigma\}, \tag{3.3}$$

where $U^{A,M}$ and $S^{A,M}$ denote the densities of inner energy and entropy of austenite and martensite, respectively. Strain $\{\varepsilon\}$ and stress $\{\sigma\}$ are written in matrix notation, which expresses the effective number of the six independent tensor coordinates [158]. Eq. 3.3 represents an idealized view, which does not take into account contributions of nonchemical energy terms, e.g. of defects or interfaces. Mechanical and thermal energy density contribute equally to the thermodynamic potentials in Eq. 3.3. This means that martensitic transformations can be affected similarly by an external mechanical load or by a temperature change.

The characteristics of the potentials of austenite and martensite are depicted schematically as a function of temperature in Fig. 3.10. At the point of intersection, both phases are in thermodynamic equilibrium. Below the thermodynamic equilibrium temperature T_0, the potential of the martensitic phase is lower than the potential of the austenitic phase. In this case, martensite is thermodynamically more stable. The same applies for the austenitic state above T_0. The difference in potentials of austenite and martensite represents the driving force for phase transformation.

Due to irreversible nonchemical contributions to the potential G, which are not considered in Eq. 3.3, martensitic transformations do not occur exactly at T_0 but in a temperature range around T_0 characterized by temperature and stress hysteresis. A main contribution is caused by friction during motion of the martensite-austenite interface [159]. Consequently, martensitic transformations require

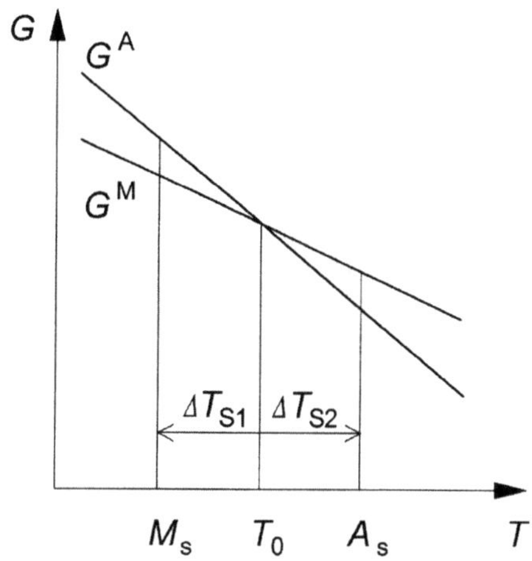

Fig. 3.10. Schematic of Gibbs potentials for martensitic and austenitic phases

supercooling by ΔT_{S1} and, accordingly, superheating by ΔT_{S2} is necessary for the reverse transformation. The equilibrium temperature T_0 can be approximated by the arithmetic mean of the temperatures M_s and A_s:

$$T_0 = \frac{M_s + A_s}{2} \ . \tag{3.4}$$

The location of T_0 is mainly determined by the chemical composition of the alloy [160].

3.3.2 Modeling

The description of martensitic transformations requires a thermodynamic model that takes into account structural changes with sufficient accuracy. Previous models can be classified by different levels of detail, as summarized in Table 3.3. Based on the model, the kinetic performance of a SMA actuator can be deduced. Complete simulation of an SMA actuator requires additional models, e.g., for the mechanical and thermal behavior.

Table 3.3. Levels of detail of models for martensitic transformations

Level of detail	Basic elements of model	Theoretical background
Microscopic	Lattice atoms, interfaces, defects	Solid state physics
Mesoscopic	Domains, grains	Solid state physics, statistical thermodynamics
Macroscopic	Volume elements, devices	Statistical thermodynamics

3.3.2.1 Microscopic Models

Microscopic models preferably take the transformation behavior of all martensitic variants into account in order to achieve a physical description of the various microscopic processes involved that is as detailed as possible. Based on the statistics of martensitic variants, the kinetic behavior of the material is deduced. Due to the high level of complexity, experimental characteristics can up until now only be reproduced qualitatively [161, 162]. As a main disadvantage, microscopic models are not technology-oriented and thus contain modeling parameters that are not measurable, e.g., the free Helmholtz energy.

3.3.2.2 Mesoscopic Models

Mesoscopic models take the mean behavior of martensitic variants in characteristic volume elements (domains or grains) and their interaction into account. One example is the one-dimensional model of Achenbach and Müller [163], which considers mesoscopic regions of austenite (A) and of two possible martensitic variants (M^+ and M^-) as basic elements of the SMA material. The fractions of both martensitic variants are determined by evolutionary equations, which are based on considerations of statistical thermodynamics and the theory of activated processes. The starting point of the model is the thermodynamic potential

$$G(\gamma) = U - ST - \tau\gamma \ , \tag{3.5}$$

which follows from Eq. 3.3 for a mesoscopic ensemble of lattice elements with the degree of freedom of one-dimensional shear γ. The potential $G(\gamma)$ can be determined by means of statistical mechanics from the potential energies of individual lattice elements.

The characteristic of $G(\gamma)$ is illustrated schematically in Fig. 3.11. Depending on the temperature of the system, $G(\gamma)$ forms up to three potential wells, which correspond to one of the three variants A, M^+, and M^-, respectively, and are separated from each other by potential barriers. The characteristic of potential is responsible for the metastable behavior of the phases and the specific properties of phase transformation. Fig. 3.11 shows schematically the characteristics of $G(\gamma)$ for the cases $T \leq M_f$ and $T \geq A_f$.

At low temperatures ($T \leq M_f$), only the potential wells of the two martensitic variants exist. For increasing temperature, an additional potential well of austenite occurs, which, in the case of $T \geq A_f$, shows a minimum below the wells of martensitic variants but is separated from the other wells by a barrier. At even higher temperatures ($T >> A_f$), the potential wells of martensitic variants disappear and austenite becomes the only stable phase.

In the presence of shear stress the potential changes by the work per volume, $\tau_0\gamma$. Thereby, the potential barrier of the martensitic variant oriented favorably with respect to the shear stress is lowered or completely eliminated. Thus, at low temperatures ($T \leq M_f$) variant coalescence is favored and at high temperatures ($T >> A_f$) a stress-induced martensitic transformation (SIM) occurs.

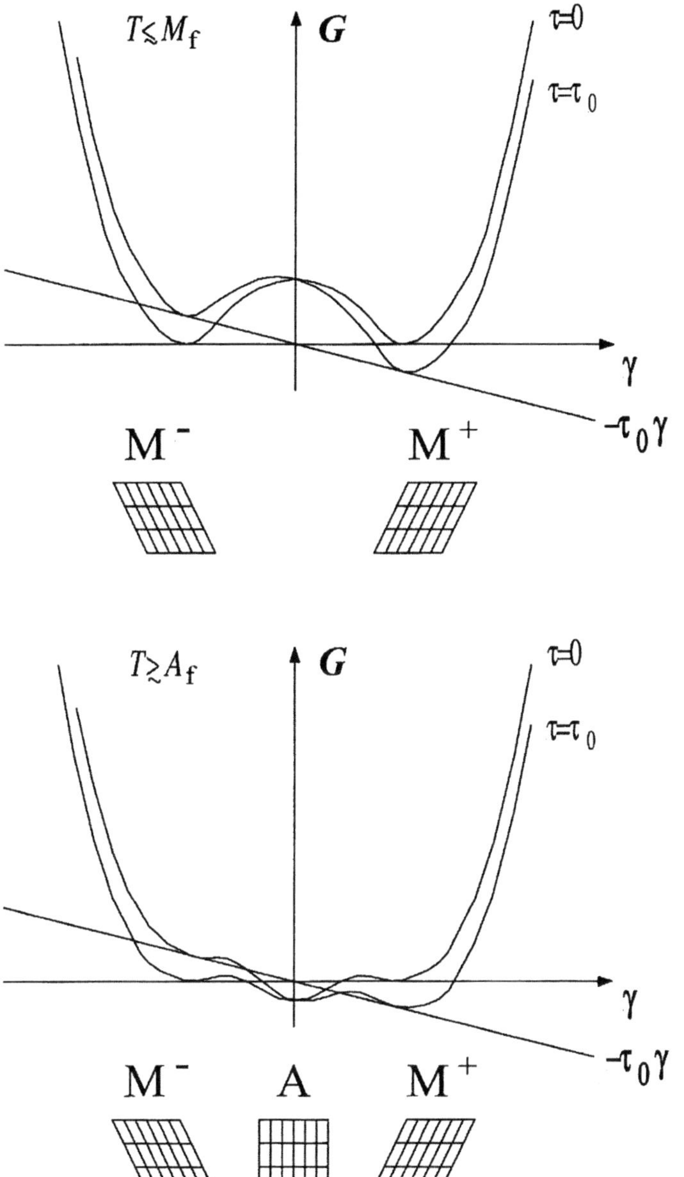

Fig. 3.11. Characteristics of the thermodynamic potential for temperatures $T \leq M_f$ (a) and $T \geq A_f$ (b) [163]. The abbreviations A, M^+, and M^- denote austenite and two martensitic variants, respectively

The shape memory behavior of simple SMA devices can be described quantitatively by the mesoscopic model outlined here [164]. However, the applicability of the model is limited, as only one dimension is considered.

3.3.2.3 Macroscopic Models

Macroscopic models consider averaged material properties in a representative volume range [165–171]. For technical applications, the use of a macroscopic model with engineering-based model parameters appears most suitable.

For SMA materials displaying a single-stage martensitic or a rhombohedral phase transformation, a two-phase macromodel will be considered, which was originally formulated by Tanaka [165]. In this model, a phenomenological macroscale constitutive law for the relation between stress, strain, temperature and phase fraction is coupled with a kinetic law describing the evolution of phase fraction. In the following, the low-temperature phase will be denoted by M. In this case, the volume fraction of M-phase is represented by an internal variable ξ, which describes both the martensitic transformation and the reorientation of martensitic variants. Thus, the volume fraction of austenite is given by $1-\xi$.

Most macroscopic models are based on recursive methods for description of the transformation process. Generally, the phase fraction ξ depends on temperature and stress. Different expressions have been developed for the kinetic behavior of ξ, which have been derived empirically [165, 166, 169] or thermodynamically [170, 172]. In the following, the kinetic behavior is expressed by exponential functions, which correspond very well to experimentally determined transformation characteristics [169]. The austenite-martensite transformation is expressed by:

$$\xi = 1 - \exp[a^M (M_s - T) + b^M \overline{\sigma}] \ . \tag{3.6}$$

The term $\overline{\sigma}$ denotes the effective stress, which is calculated from the cartesian stress components. The constants a^M and b^M are determined experimentally. Assuming that the transformation is finished for $\xi = 0.99$, the constant a^M is given by $a^M = \ln(0.01)\,(M_s - M_f)^{-1}$. The reverse transformation is expressed accordingly [169]:

$$\xi = \exp[a^A (A_s - T) + b^A \overline{\sigma}] \ , \tag{3.7}$$

where the constant a^A is given by $a^A = \ln(0.01)\,(A_s - A_f)^{-1}$. By Eqs. (3.6) and (3.7), the evolution of complete hysteresis loops can be adequately described.

The constants $b^M/a^M=1/c^M$ and $b^A/a^A=1/c^A$ describe the linear dependence of transformation temperatures on the effective stress $\overline{\sigma}$ usually denoted as stress rate. The stress rate follows from the Clausius-Clapeyron relation for the martensitic transformation. By assuming the equilibrium condition $dG^A=dG^M$, the following expression is obtained from Eq. 3.3 in a one-dimensional case:

$$0 = -(S^M - S^A)\,\mathrm{d}T - (\varepsilon^M - \varepsilon^A)\,\mathrm{d}\sigma \ . \tag{3.8}$$

After some conversions, the Clausius-Clapeyron equation is obtained in the following form [173]:

$$\frac{\mathrm{d}\sigma}{\mathrm{d}T} = -\frac{\Delta S}{\varepsilon^{tr}} = -\frac{\Delta H \rho}{\varepsilon^{tr} T_0} = c \qquad (3.9)$$

In this case, it has been assumed that the elastic strains in austenite and martensite are identical. The terms ε^{tr}, ΔH, and ρ denote transformation strain, transformation enthalpy, and mass density, respectively. The change in entropy density occuring at the transformation is given by $\Delta S = S^M - S^A$. In the three-dimensional case, it is assumed that the shape memory effect shows a von Mises response. Thus, the one-dimensional stress σ and transformation strain ε^{tr} can be replaced by the effective terms $\bar{\sigma}$ and $\bar{\varepsilon}^{tr}$ [169].

The transformation enthalpy ΔH can be determined experimentally, e.g., by differential scanning calorimetry (DSC). For the martensite-austenite reverse transformation in NiTi, ΔH is, for instance, 15.4 J/g [174]. The corresponding stress rate at a transformation temperature of 300 K and an effective transformation strain of 5% is 6.6 MPa/K, which agrees well with measurements on polycrystalline NiTi [174].

The stress-temperature regimes of martensitic transformation are illustrated in Fig. 3.12. In the regime $T<M_s$, martensite is formed exclusively by thermally

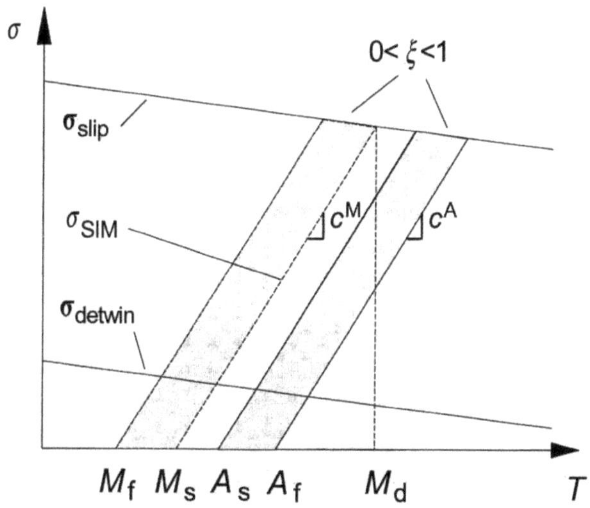

Fig. 3.12. Stress-temperature phase diagram: The straight lines σ_{detwin}, σ_{SIM}, and σ_{slip} denote the threshold stress for the processes of variant coalescence, stress-induced martensite formation, and slip deformation. The stress-temperature regimes of martensitic transformation are indicated. The slopes c^M and c^A are assumed to be identical

induced transformation. At temperatures $T>M_s$, oriented martensitic variants can be generated directly by SIM formation. The threshold stress required for this process increases linearly with temperature, as described by Eq. 3.9. Thus, a linear shift of the stress-temperature regimes of forward and reverse transformation results. In stress-free condition, martensite is only thermodynamically stable at temperatures $T < A_f$. However, in the presence of stress, martensite is stabilized up to the critical temperature M_d, above which the critical stress for slip deformation σ_{slip} is exceeded.

Eqs. 3.6 and 3.7 are used as constitutive equations for the calculation of the kinetic behavior of SMAs. The presented model captures the basic features of transformation kinetics, which is sufficient for various applications. However, depending on the problem, model extensions are required to improve accuracy. In the following, several aspects of model extensions are briefly addressed:

(1) In partial phase transformations, the transformation characteristic of the actual load case depends on the phase fractions remaining from previous load cases. Partial hysteresis loops can be described by memory terms for forward and reverse transformation $\xi_{memory}^{\uparrow,\downarrow}$ in the kinetic equations [175].

(2) Periodic temperature and stress load may be associated with training effects and irreversible processes, which cause a change in mechanical characteristics, particularly during the first few load cycles until a stationary performance appears. Such effects remaining from previous load cycles can be taken into account similarly to (1) by introducing further memory terms.

(3) By differentiating between self-accommodated and oriented martensitic variants, the mechanisms of martensite formation are taken into account more accurately and, thus, greater accuracy in the description of phase transformation may be achieved [167]. As shown in Fig. 3.9, thermally induced transformation at zero mechanical load is associated with the formation of self-accommodated (sa) martensitic variants, while the presence of stress causes the formation of oriented (o) martensitic variants and/or the transformation of martensitic variants into each other. These mechanisms can be taken into account by introducing two internal variables, ξ_{sa} and ξ_o, with different kinetic behavior. In this case, the volume fraction of austenite becomes $1-\xi_{sa}-\xi_o$.

(4) Tanaka-type models treat each loading direction in the same way. For complex loading cases such as non-proportional loading, improved models have to be considered that take coupling between formation and annihilation of martensitic variants of different loading directions and reorientation of stress-induced martensitic variants into account [176, 177].

As a matter of principle, macroscopic models have only a limited significance due to the use of averaged material parameters. Empirical material data represent a special microstructure that, strictly speaking, may only be applied to the material being experimentally investigated. Taking this restriction into account, the use of a Tanaka-type model for SMA microactuators can lead to satisfactory agreement with experimental performance data, as will be discussed in Chaps. 5 and 6.

3.3.2.4 Simulation of SMA Components

For simulation of SMA components, the mechanical and thermal properties have to be considered in addition to the kinetic behavior. As the components can have an arbitrary three-dimensional shape, field simulations may be required. In the following, models for simulation of the mechanical and thermal properties of SMA components will be presented.

Mechanical Model: For a given load, the mechanical properties of an SMA component depend on the geometry, as well as on the microstructure. In particular, the stress-strain relation depends on the phase fractions of austenite and martensite. In the following, parallel coupling of both phases is assumed according to the Voigt model of micromechanics, which is illustrated in Fig. 3.13. Plastic strain contributions caused by martensitic phase transformation or by reorientation of martensitic variants are coupled serially to the total strain. The total stress of the austenite-martensite compound is described by a rule of mixture:

$$\{\sigma\} = (1-\xi)\{\sigma\}^A + \xi\{\sigma\}^M \tag{3.10}$$

The corresponding elastic strains $\{\varepsilon\}^{el}, \{\varepsilon\}^A_{el}$, and $\{\varepsilon\}^M_{el}$ are identical due to parallel coupling:

$$\{\varepsilon\}^M_{el} = \{\varepsilon\}^A_{el} = \{\varepsilon\}^{el} \ . \tag{3.11}$$

As is common for elastic material behavior, Hooke's law holds between stress and strain:

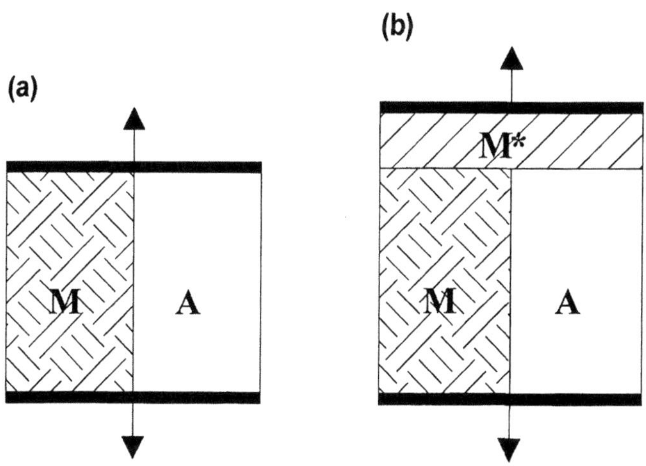

Fig. 3.13. Voigt model for SMAs: (a) parallel coupling of strains from austenitic and martensitic phases, (b) serial coupling of strains originating from transformed or reoriented martensite denoted as M*

$$\{\sigma\}^A = [E^A]\{\varepsilon\}_{el}^A \tag{3.12}$$

$$\{\sigma\}^M = [E^M]\{\varepsilon\}_{el}^M \ , \tag{3.13}$$

where $[E^A]$ and $[E^M]$ are the matrices of E-moduli in austenitic and martensitic conditions.

From Eqs. 3.10 – 3.13 the relationship between stress and elastic strain follows:

$$\{\sigma\} = ((1-\xi)[E^A] + \xi[E^M])\{\varepsilon\}^{el} \ . \tag{3.14}$$

Assuming furthermore that the total strain is the sum of elastic and transformation strain:

$$\{\varepsilon\} = \{\varepsilon\}^{el} + \{\varepsilon\}^{tr} \ , \tag{3.15}$$

Eq. 3.14 leads to

$$\{\sigma\} = ((1-\xi)[E^A] + \xi[E^M])(\{\varepsilon\} - \{\varepsilon\}^{tr}) \,. \tag{3.16}$$

The term $\{\varepsilon\}^{tr}$ denotes the matrix of nonlinear thermoelastic strains of phase transformation. Eq. 3.16 may be used as a constitutive equation for the calculation of the three-dimensional mechanical behavior of SMA components. Other models assume a serial coupling of both austenitic and martensitic phases according to the Reuss model of micromechanics and, thus, obtain different approximations for the stress-strain relation of the microstructure [170, 171].

The effect of thermal expansion may be taken into account by an additional term in Eq. 3.16:

$$\{\sigma\}_{therm} = \{\alpha'\}(T - T') \ , \tag{3.17}$$

where $\{\alpha'\}$ is the matrix of thermal expansion coefficients consisting of the contributions of austenite and martensite weighted by the corresponding phase fractions. The temperature T' is the temperature of zero thermal strain.

For simulation of SMA actuators, the mechanical behavior of other components besides the SMA component has to be considered, in particular, of the reset mechanism. This topic will be addressed in Chap. 4.

Thermal model: The temperature profile in an SMA component is determined by heat exchange with its environment, which depends on the boundary conditions of temperature and stress load. Assuming an isotropic material, the heat transfer problem within a three-dimensional body is defined by the differential equation:

$$m(c_p + h(T))\dot{T} = \dot{Q} - \lambda\Delta T \ , \tag{3.18}$$

which can be used to calculate the temperature T in Eqs. 3.6, 3.7, and 3.17. In Eq. 3.18 the following contributions are taken into account:

- the specific heat of the SMA component, whereby c_p denotes the specific heat capacity at constant pressure,

- the latent heat of phase transformation *h(T)*, which is released upon cooling and has to be supplied upon heating,
- the heating power $\dot{Q}$ introduced into the system from the outside, which is generated, e.g., by electrical current, and
- the heat exchange with the environment by heat conduction, where λ denotes the thermal conductivity.

The parameter *m* is the total mass of the SMA component. For applications in the temperature range below 200 °C, the effect of heat radiation can be neglected.

At the surface of the SMA component, the Cauchy boundary condition holds:

$$-\lambda\vec{\nabla}T = K(T - T_E) \ , \tag{3.19}$$

which takes into account the convective heat exchange with the environment of temperature T_E. The parameter *K* denotes the heat-transfer coefficient.

3.4 NiTi-Based Alloys

From today's point of view, the Ni-Ti-Cu-Pd- material system possesses the largest application potential as compared to other shape memory alloys. Comprehensive studies, in particular of NiTi wires, have led to an extensive unterstanding of this alloy, e.g., in terms of the effects of stoichiometry, thermomechanical treatment, and chemical composition [139, 178]. In contrast, the specific properties of rolled NiTi foils or sputtered thin films are still the subjects of present research activities.

3.4.1 Metallurgy

NiTi is an ordered intermetallic compound. According to the phase diagram, the NiTi compound exists as a stable phase down to room temperature. However, the stoichiometric range is rather narrow at low temperatures. Below 923 K, the B2-phase region only ranges from 50.0 to 50.5 at.% Ni [139]. For this reason, the alloys frequently contain precipitates of a second intermetallic phase such as Ti_3Ni_4 or $TiNi_3$. Moreover, the presence of oxygen results in Ti-rich oxide compounds, e.g., Ti_4Ni_2O, which cause the Ti content in the matrix to decrease and the stoichiometric range to be reduced even further [179].

Ni-rich and Ti-rich alloys have characteristic differences. Ni-rich alloys exhibit a strong dependence of the M_s temperature on the chemical composition. To keep the temperature variation below 2 K, precise control of the Ni content of at least 0.05 at.% is required during melting. This dependence is much smaller for Ti-rich alloys, as the formation of Ti-rich precipitates ensures a hardly changed matrix composition. The M_s temperature of Ni-rich alloys is below room temperature. Maximum M_s temperatures between 90 °C and 100 °C are reached at a Ni content below 49.5 at.%.

At a Ni content in excess of 50.5 at.%, the alloys tend to behave pseudoelastically in a certain temperature range above the M_s temperature. In fully annealed and quenched alloys, the pseudoelastic temperature range shifts toward lower temperatures with increasing Ni content. For heat treatment temperatures between 300 and 550 °C, Ti_3Ni_4 precipitates may form, which cause the Ni content of the matrix to decrease, and, hence, the pseudoelastic temperature range to increase.

Alloys having a Ni content below 50.5 at.% are not sensitive to a heat treatment below the recrystallization temperature of about 550 °C, as no Ti_3Ni_4 precipitates form in this case. Still, their properties are strongly affected by a thermomechanical treatment (cold deformation and subsequent heat treatment below the recrystallization temperature). It is assumed that the dislocations generated by cold deformation reorient to cells during thermal treatment. Thus, dislocation-depleted zones develop, where movements of the twin boundaries are possible. Cell formation results in material hardening, while the tendency toward plastic deformation is suppressed. The degree of cold deformation and the temperature of heat treatment influence the plastic deformation limit and the achievable reversible strain in an opposite manner and, hence, have to be selected carefully to obtain the desired material properties. By a thermomechanical treatment of polycrystalline NiTi, a reversible strain of 7 to 8% can be reached [154]. Above the recrystallization temperature, the dislocation density generated by cold deformation is reduced again, which causes the shape memory properties to deteriorate.

Studies with regard to the influence of the grain size in polycrystalline shape memory alloys on the shape memory properties also reveal an increase in reversible strain with decreasing grain size [180]. This property is of interest particularly for sputtered thin films.

3.4.2 Phase Transformations

Ti-rich and completely annealed equiatomic alloys exhibit a single-stage martensitic transformation from the B2-phase directly to the monoclinic B19'-phase. In Ni-rich alloys and thermomechanically treated equiatomic alloys, however, a two-stage transformation with the sequence

$$\text{B2-phase} \rightarrow \text{R-phase} \rightarrow \text{B19'-phase}$$

is observed upon cooling. The rhombohedral (R-)phase possesses a trigonal crystal structure and develops from the B2-phase by lattice deformation in the [111]-direction [140]. In the [111]-direction, a maximum reversible lattice strain of about 1% is reached. In polycrystalline materials, maximum reversible lattice strains range from 0.5 to 0.8%. Occurrence of the R-phase is reflected by additional X-ray reflexes at the 1/3-positions of the reciprocal lattice [181], which originally were interpreted as pre-martensitic phenomena.

Fig. 3.14 shows the transformation sequence of a cantilever beam specimen made of thermomechanically treated equiatomic NiTi in displacement and electrical resistance. The vertical displacement of the cantilever beam has been recorded under constant load applied at the front end of the beam.

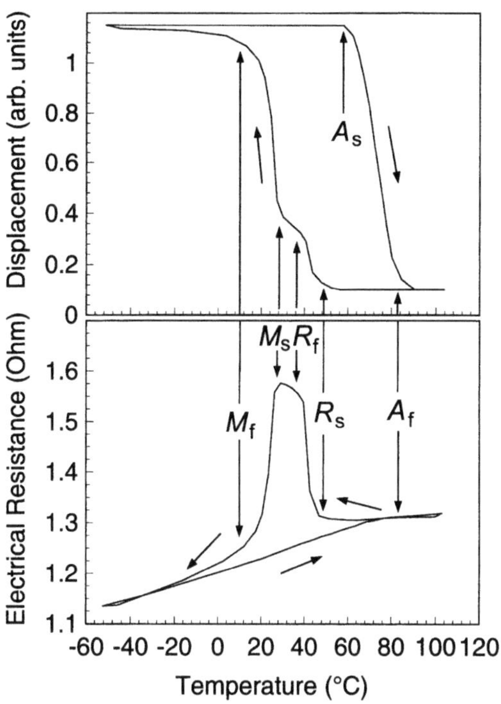

Fig. 3.14. Displacement and electrical resistance characteristic of a NiTi cantilever beam specimen with a Ti content of 50 at.% (cold deformation 20%, heat treatment at 823 K for 10 min) [182]

Upon cooling of the specimen, the electrical resistance starts to increase at the rhombohedral start temperature R_s. This temperature corresponds to the start temperature of displacement of the cantilever beam. In the temperature range between R_s and the finish temperature R_f, both B2- and R-phases coexist. The temperature at which the electrical resistance decreases coincides with the second steep slope of beam displacement. At this point, the M_s temperature is reached at which the transformation R $\rightarrow$ B19' starts. Below the M_s temperature, R- and B19'-phases coexist until the M_f temperature is reached. During heating, only a single-stage transformation from B19'-phase directly to B2-phase takes place.

The result of a calorimetric measurement using the DSC (differential scanning calorimetry) method is presented in Fig. 3.15 for comparison. Here, the same transformations occur. However, the transformation temperatures are shifted toward somewhat smaller values compared to the mechanical measurements, because no external load is applied. If cooling is interrupted between the temperatures M_s and M_f and subsequent heating takes place, a two-stage transformation can be observed during heating as well. When cycling the temperature from the

B2-phase up to the M_s temperature, only a single-stage R-phase transformation is observed.

As an example, the mechanical and electrical behavior of a cantilever beam when heated electrically above the M_s temperature is shown in Fig. 3.16. Small hysteresis widths occur, which correspond to a temperature difference of about 1 K. This behavior results from the small transformation strain below 0.8% [140]. In addition to the thermally induced shape memory effect, the R-phase transformation also exhibits pseudoelastic behavior [140].

Although reversible transformation strain is 0.8% at the maximum, an R-phase transformation may be very useful for actuator applications. It exhibits an abrupt thermal response. When used for position control, the small hysteresis allows for high precision without any sophisticated control scheme. Furthermore, the shape memory effect is stable over many transformation cycles, see Sect. 3.4.3.4.

3.4.3 Mechanical Behavior

The mechanical behavior of a shape memory alloy depends on the type of thermomechanical loading. Material investigations are usually performed for one of the two boundary cases of constant temperature or constant force. Real load cases in microactuation are generally characterized by a combination of stress and temperature loading.

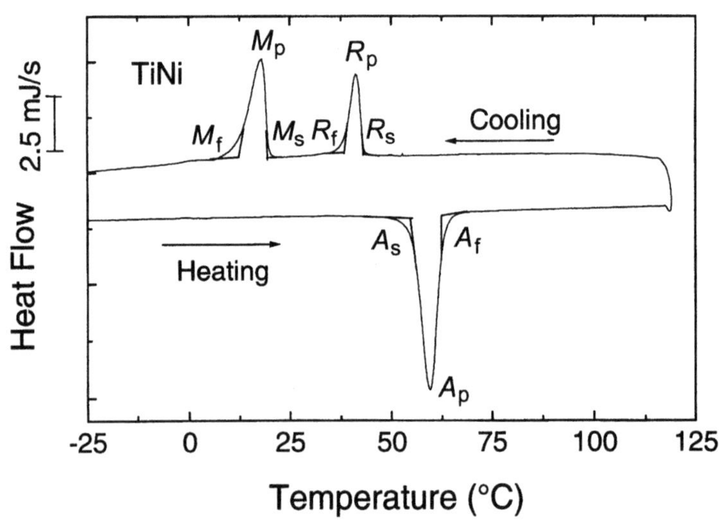

Fig. 3.15. DSC measurement of a NiTi alloy with a Ti content of 50 at.% (cold deformation 20%, heat treatment at 823 K for 10 min) [183]

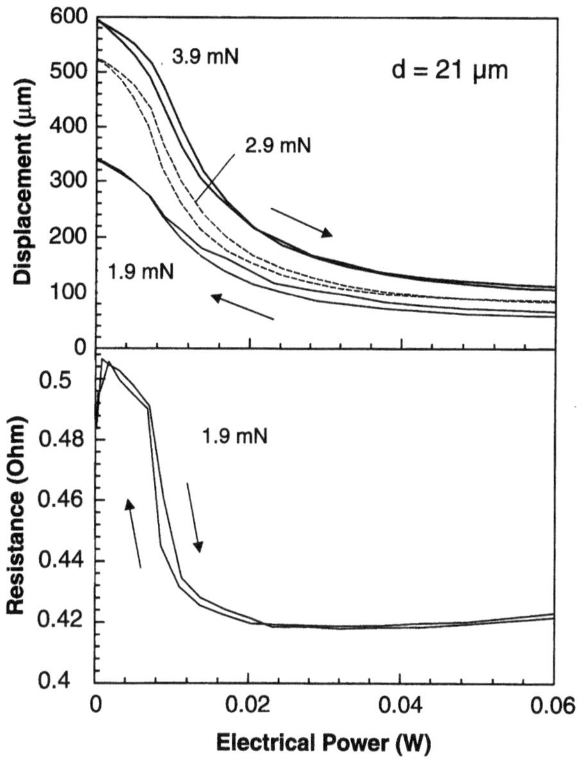

Fig. 3.16. Displacement and electrical resistance of a NiTi cantilever beam specimen with a Ti content of 50.3 at.% (cold deformation 30%, heat treatment at 673 K for 1 h) [184]

3.4.3.1 Behavior at Constant Temperature

The behavior of the stress-strain characteristics depends on the temperature in the material. A survey of typical stress-strain characteristics is given in Fig. 3.17 [140].

In the cases of R-phase and martensitic phase transformation, a total of six temperature ranges can be distinguished:

(a) $T < M_f$: Single-stage transformation associated with coalescence of martensite variants.

(b) $M_f < T < M_s$: Two-stage transformation caused by the coexistence of R-phase and martensite. Each stage is associated with coalescence of R-phase variants and martensite variants.

(c) $M_s < T < A_f$: Deformation first starts with the coalescence of R-phase variants. Upon further loading, stress-induced martensite develops.

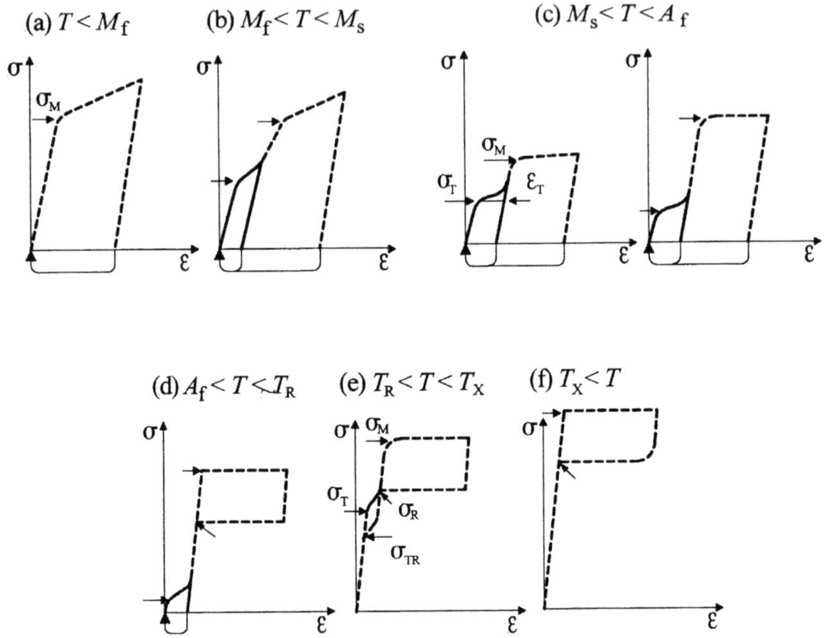

Fig. 3.17. Classification of isothermal stress-strain characteristics of a NiTi alloy, showing an R-phase and a martensitic phase transformation [140]

(d) $A_f < T < T_R$: Pseudoelastic range due to a martensitic transformation. Part of the deformation, however, is caused by coalescence of R-phase variants. The end temperature of the R-phase reverse transformation is referred to as T_R.

(e) $T_R < T < T_X$: Two-stage pseudoelasticity caused by R-phase and martensitic transformation. As the critical stress for generating the R-phase increases more rapidly with temperature than in the case of the martensitic phase, the critical stresses intersect at the temperature T_X.

(f) $T > T_X$: Single-stage pseudoelasticity caused by a martensitic transformation, as the critical stress is too high for R-phase transformation.

A change in the type of loading (tension, bending, torsion, compression) results in corresponding changes of slopes of the stress-strain characteristics and of hysteresis widths within the temperature ranges.

3.4.3.2 Behavior at Constant Force

Fig. 3.18 shows a series of displacement-temperature characteristics recorded during a bending test of a cantilever beam specimen made of a thermomechanically treated NiTi plate with a Ti fraction of 50.3 at.% [185]. Upon cooling, the material first transforms to the R-phase, which is not resolved in the displacement characteristics. The existence of R-phase can, however, be concluded from corresponding resistance characteristics. Upon heating, a direct B2 → B19' transformation takes place. With increasing load, the transformation temperatures rise as a result of the external stress applied. Bending of the beam is determined mainly by the maximum stress occuring near the beam suspension. Hence, the temperature shift can be correlated to the maximum stress in the beam [185]. The resulting relationship between maximum stress and transformation temperatures is presented in Fig. 3.19.

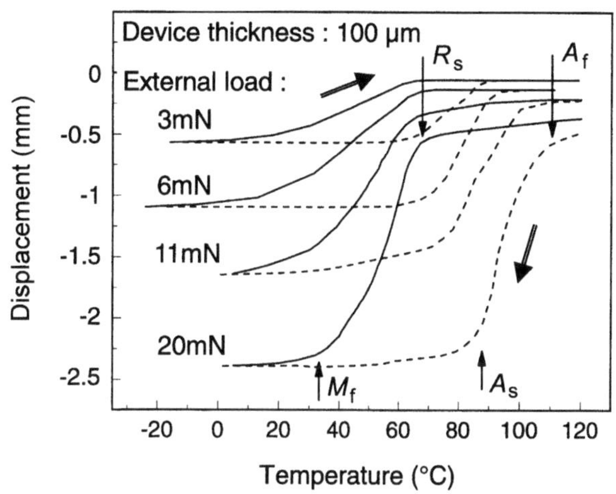

Fig. 3.18. Displacement-temperature characteristics of a NiTi cantilever beam specimen with a Ti content of 50.3 at.% (cold deformation 30%, heat treatment at 673 K for 1 h) [185]

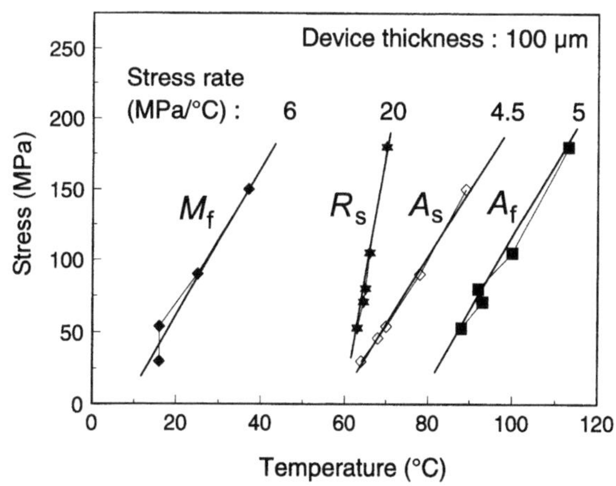

Fig. 3.19. Stress rates determined from bending tests, as shown in Fig. 3.18 [185]

A linear relationship is obtained as predicted by the Clausius-Clapeyron equation, see Eq. 3.9. While typical stress rates between 4 and 6 MPa/K are determined for the martensitic transformation temperatures M_f, A_s, and A_f, the stress rate of the R_s temperature is about 20 MPa/K. The considerably smaller stress dependence of the R_s temperature results from the smaller transformation strain of the B2 → R transformation [140].

3.4.3.3 Orientation Dependence

Polycrystalline material consists of statistically distributed grains that cause isotropic shape memory properties. By cold rolling, however, the material is given a preferential orientation, which arises from sliding planes being favored during the plastic deformation of the plates. The resulting anisotropy effects may be important for the design of SMA microactuators and have been, therefore, investigated systematically [19, 184, 186–189]. Fig. 3.20 shows the result of tensile tests of rolled NiTi plates with the tensile axis being along the rolling direction and in transverse direction, respectively [186].

The maximum reversible strain ε_A during reverse transformation to austenite first displays a linear increase as a function of stress. It reaches an orientation-dependent maximum of 5% for tensile loading along the rolling direction and about 4.1% for loading in the transverse direction. This corresponds to a maximum anisotropy of 20%. The higher strain in the rolling direction may be attributed to a preferential orientation in the [011]- and [111]-directions [187]. Critical stress for plastic deformation is about 90 MPa in the case of tensile loading in the transverse direction and about 60 MPa in the case of tensile loading along the rolling direction.

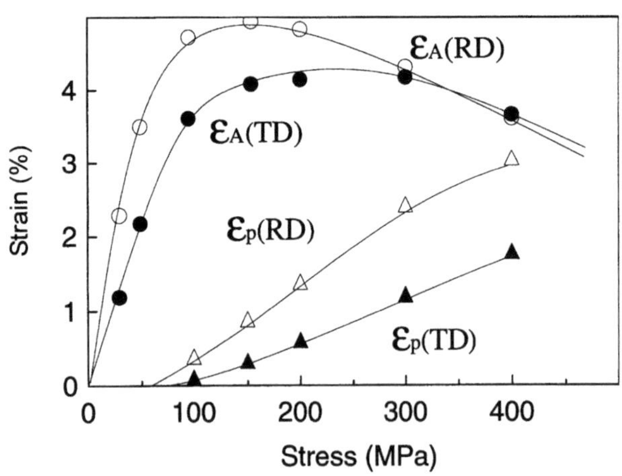

Fig. 3.20. Maximum reversible strain ε_A and plastic strain ε_P of a rolled NiTi plate with a Ti content of 50 at.% (cold deformation 20%, heat treatment at 823 K for 10 min) for tensile axis along rolling direction (RD) and transverse direction (TD), respectively [186]

3.4.3.4 Fatigue Behavior

Dependence of the achievable reversible strain on the maximum stress, maximum strain, and the number of transformation cycles is of considerable interest with regard to the service life of SMA microactuators. As explained in Sect. 3.2, the achievable reversible strain of a certain alloy system and a certain transformation type is determined by the crystal structure, including lattice defects.

Table 3.4 shows empirically determined numbers of reversible cycles of martensitic transformation as a function of the maximum stress and strain of thermomechanically treated polycrystalline NiTi alloys [190]. When using the maximum possible strain in martensitic condition of 7–8%, the transformation cycle is passed once in a reversible manner under optimum conditions. The same applies to a maximum stress load of about 500 MPa. Repeated cycling of stress or temperature causes irreversible strains in this case. Thus, the maximum reversible cycle number given by the application determines the technically possible load limit. Consequently, in most applications the technically usable strain or stress range is far below 8% or 500 MPa, respectively.

The data given in Table 3.4 can be considered rough orientation values for the application of thermomechanically treated NiTi alloys in microactuators. Detailed studies of the fatigue behavior of NiTi are given in the literature [139, 191].

If the work production is based on an R-phase transformation, the usable strain is limited to a maximum of about 0.8% [140]. However, in this case no fatigue effects are observed even at high cycle numbers above 10^6.

Table 3.4. Achievable number of reversible transformation cycles as a function of maximum strain and stress [190].

Cycles	Maximum strain	Maximum stress
1	8%	500 MPa
100	4%	275 MPa
10,000	2%	140 MPa
>100,000	1%	70 MPa

3.4.4 Ternary Alloys on the Basis of NiTi

Alloying additional elements to the NiTi alloy system influences various material properties such as transformation temperatures, hysteresis widths, material hardness, etc. and allows suppressing the R-phase transformation. NiTiCu and NiTiPd alloys are of particular interest to the microactuators described in Chaps. 5 and 6.

3.4.4.1 NiTiCu Alloys

Material properties of NiTiCu alloys were studied in a number of publications, see, e.g., [139, 192]. In NiTi alloys, Ni atoms may be replaced by Cu atoms by up to about 30 at.% without modifying the austenitic high-temperature phase. Pure CuTi, however, possesses a tetragonal structure. Consequently, a phase mixture of cubic and tetragonal phases occurs for a Cu content in excess of 30 at.%. Starting from a Cu content of 10–15 at.%, the martensitic low-temperature phase changes from a monoclinic (B19') to an orthorhombic structure (B19) without considerable change of the shape memory properties.

As an example, Fig. 3.21 shows the transformation behavior of a NiTiCu cantilever beam specimen of 100 μm thickness, as reflected by the electrical resistance and mechanical displacement [193]. The Cu content of the specimen is 5.9 at.%. The vertical displacement of the cantilever beam was determined under constant load applied at the front end of the beam. A single-stage transformation is observed. The transformation takes place above room temperature showing a typical hysteresis width below 20 K.

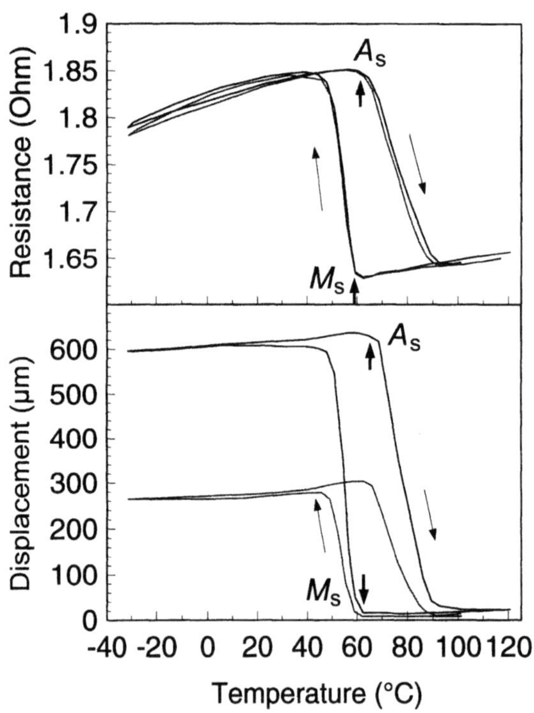

Fig. 3.21. Electrical resistance and displacement of a cantilever beam specimen of NiTiCu with a Cu content of 5.9 at.% (cold deformation 25%, heat treatment at 803 K for 15 min) [193]

The major material properties observed for a Cu content below 30 at.% may be summarized as follows:

- At a Cu content below 5 at.%, a two-stage transformation via the R-phase occurs: B2 → R → B19'.
- For a Cu content in the range of 5 and 15 at.%, phase transformation takes place in two stages: B2 → B19 → B19'. Above 15 at.%, only a single-stage transformation to the orthorhombic structure occurs: B2 → B19.
- Hysteresis widths in NiTiCu alloys are much smaller than in NiTi alloys. At a Cu content of 5 at.%, the hysteresis width is already reduced from about 30 K to about 15 K. A further increase to 25 at.% leads to an additional slight decrease to about 10 K. The significance for actuator applications lies in, e.g., faster response times and a reduced effort for position control.
- The occurrence of smaller hysteresis widths is closely associated with the decrease of the elastic limit of martensite. As described in Sect. 3.2.2, the elastic limit in the martensitic condition is determined by the mobility of twin boundaries. At a Cu content of 10 at.%, the value decreases to about half the value of NiTi. This behavior is advantageous to actuator applications, as a small force is required for resetting the actuator in martensitic condition, which results in a higher work per actuation cycle.
- Irrespective of the Cu concentration, the M_s temperature remains nearly constant. The high sensitivity of the M_s temperature to changes in the Ni/Ti ratio in binary alloys is suppressed by the addition of Cu atoms. Thus, controllability and reproducibility of the M_s temperature are improved during material production.
- The maximum reversible strain decreases for increasing Cu content. At a load of 275 MPa, it amounts to about 3% in NiTiCu with a Cu content of 10 at.%. The corresponding value reached in NiTi, however, is about 4.8%.
- Above a Cu content of 10 at.%, an increase in brittleness is observed.
- In NiTiCu, the electrical resistance in martensitic condition is higher than in austenitic condition. In NiTi, the opposite is the case.

3.4.4.2 NiTiPd Alloys

In past years, NiTiPd alloys gained increasing attention [194]. This is mainly due to the strong increase in transformation temperatures with increasing Pd content, which is of considerable interest to actuator applications. By replacing Ni atoms with Pd atoms, the M_s temperature can be shifted from about 50°C for NiTi to 540°C for TiPd [195]. At small Pd concentrations, the M_s temperature is reduced first.

The major material properties may be summarized as follows:

- For a Pd content below about 25 at.%, a two-stage transformation via the R-phase occurs: B2 → R → B19'.
- Above a Pd content of about 25 at.%, a single-stage transformation to the orthorhombic structure takes place: B2 → B19.
- The hysteresis widths are not changed significantly with increasing Pd content.

- The maximum reversible strain decreases with increasing Pd content. Below a Pd content of 30 at.%, maximum reversible strains of up to 4.5% are observed. Above a Pd content of 45 at.%, the maximum reversible strain is below 1%.
- NiTiPd alloys exhibit an increased brittleness. This adversely affects their workability, e.g., by cold rolling.

3.4.5 Thin Films

SMA thin films in the thickness range below 20 μm are particularly suited for the realization of microactuators. The fabrication of NiTi-based thin films has been accomplished using, e.g., magnetron sputtering [14–18], ion beam sputtering [13], and laser ablation [196]. The most frequently used fabrication method so far is magnetron sputtering.

3.4.5.1 Binary NiTi Alloys

The chemical composition of the thin films is adjusted in magnetron sputtering by the composition of the sputtering target. When using alloy targets, a decrease of the Ti content and an increase of the Ni content are observed in the crystal matrix, which has not yet been clarified completely. A possible reason might be oxidation of Ti during deposition. This effect can be compensated for, e.g., by partially covering the alloy target with pure Ti pieces. At substrate temperatures below 200°C, the deposited films are amorphous, such that no shape memory effect can be observed. In this case, an annealing treatment above the crystallization temperature of about 400°C is required. Alternatively, crystalline thin films may be fabricated by respective heating of the substrate.

Crystallization takes place by nucleation and growth of grains having a B2 structure. By this process, Ti_3Ni_4 precipitates form in Ni-rich thin films with a Ni content higher than 50.6 at.%. If the crystallization of the thin films is performed in a rolled state, precipitates with a certain preferential orientation develop due to the existing stress field. They act as an intrinsic counterspring and thus cause the all-round effect [143]. In the case of a nearly equiatomic composition with a Ni content between 50 and 50.5 at.%, no precipitation takes place. In Ti-rich thin films, Guinier-Preston zones and Ti_2Ni precipitates may form depending on the annealing parameters [197].

The achievable shape memory properties do not differ significantly from the properties of bulk materials produced by melting metallurgy [16]. In sputtered and crystallized films, however, much smaller grain sizes are formed depending on the temperature and duration of the heat treatment. Consequently, the crystal matrix is strengthened with regard to its plastic deformability and, hence, the shape memory behavior is improved. A comparable effect is observed for the formation of coherent Guinier-Preston zones. In this case, maximum reversible strains of 5.5% can be reached at a stress of 240 MPa [197]. Thermomechanical processes for the stabilization of shape memory effects, which have proven suitable for bulk materials produced by melting metallurgy, may not be transferred to thin films. And, vice versa, the process of Guinier-Preston zone formation, which is suited for thin films, may not be applied to bulk materials, as this would require an amorphous initial state [198].

The typical behavior of the resistance and displacement of a cantilever beam specimen made of a sputtered NiTi thin film of 8 μm thickness is shown in Fig. 3.22 [199]. Above room temperature, an R-phase transformation takes place at the temperatures R_s/R_f of 324/306 K. The hysteresis width of about 1 K is characteristic of an R-phase transformation. This property is of particular interest to microactuator applications. In particular, the similar transformation behavior of displacement and resistance results in an almost linear relationship with respect to each other [200]. This allows for smart actuation of components made of this material by using the intrinsic sensing and actuation properties simultaneously [201].

3.4.5.2 NiTiCu Alloys

Studies with regard to the shape memory effect in NiTiCu thin films with Cu content below 9.5 at.% reveal a maximum reversible strain of 3.9% at a stress of

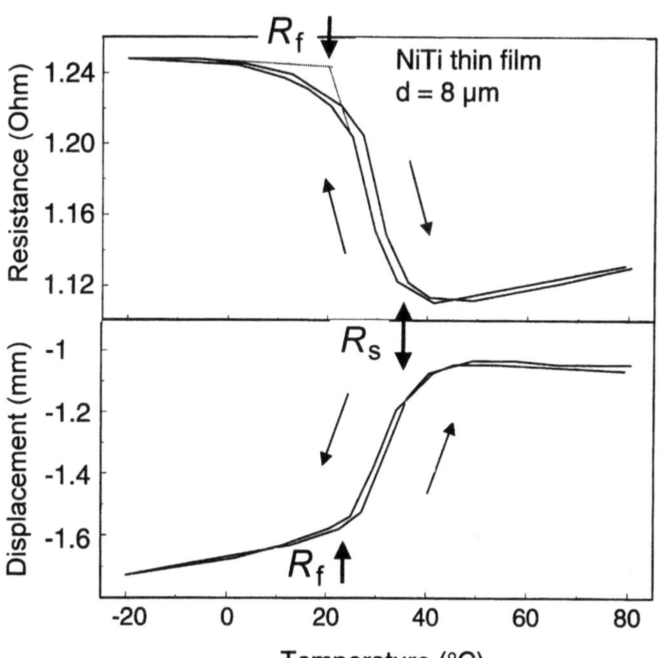

Fig. 3.22. Displacement and electrical resistance of a cantilever beam specimen made of a sputtered NiTi thin film (integral chemical composition: $Ni_{48}Ti_{52}$, sputtering power 300 W, annealing: 843 K – 2.7 ks) [200]

about 200 MPa [202]. Up to a Cu content of 5 at.%, the phase transformation occurs in a single stage: B2 → B19'. At 9.5 at.%, the phase transformation is accomplished in two stages: B2 → B19 → B19'. When increasing the Cu content to 18 at.%, a single stage transformation of B2 → B19 is observed, as in bulk materials. In this case, the maximum reversible strain drops to 1.1%.

In Ti-rich NiTiCu alloys, coherent Guinier-Preston zones may be formed by optimizing the annealing parameters, as in the case of binary NiTi thin films, which improve the shape memory behavior [198]. In this case, a maximum reversible strain of 5.2% could be achieved at a stress of 470 MPa in a NiTiCu thin film with a Ti content of 50.8 at.% and a Cu content of 6.2 at.%.

For increasing Cu content, the hysteresis width drops from about 40 K to a minimum of 18 K [198, 202, 203]. Hence, the minimum hysteresis width achievable is higher than for comparable bulk materials [192].

Fig. 3.23 shows an electrical resistance characteristic of a NiTiCu cantilever beam specimen of 10 μm thickness. A single-stage martensitic transformation with a hysteresis width of 18 K is observed. The NiTiCu thin film serves as a starting material for development of the microvalves presented in Sect. 5.2.3.

3.4.5.3 NiTiPd Alloys

Only a few studies on NiTiPd thin films have been published thus far [204, 205]. Fig. 3.24 shows DSC characteristics of sputtered NiTiPd thin films with a Pd content of 20 and 60 at.%, respectively [204]. When increasing the Pd content, trans-

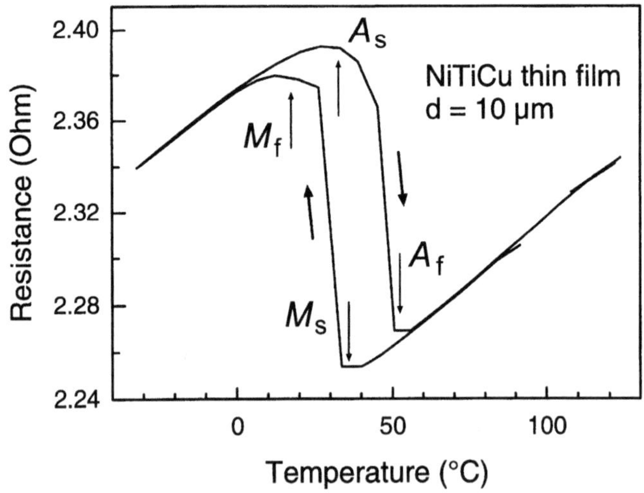

Fig. 3.23. Electrical resistance of a cantilever beam specimen made of a sputtered NiTiCu thin film (integral chemical composition: $Ni_{30}Ti_{53}Cu_{15}$, sputtering power: 300 W, annealing treatment: 843 K – 2.7 ks)

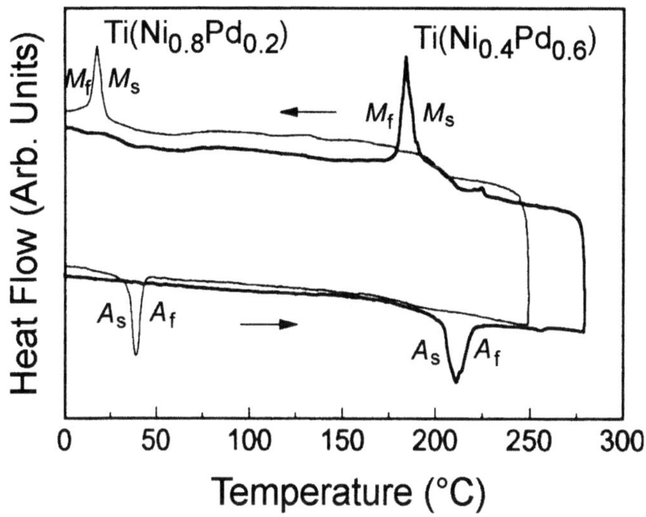

Fig. 3.24. DSC characteristics of sputtered NiTiPd thin films with different Pd content (sputtering power: 300 W, annealing treatment: 1023 K – 3.6. ks) [204]

formation temperatures rise considerably, as for bulk materials. In binary NiPd thin films, the transformation temperatures M_f /A_f of 498/570 °C are reached eventually [204].

Studies of the maximum reversible strain in a thin film of the chemical composition $Ni_{26.4}Ti_{51.8}Pd_{21.8}$ yield a value of 1.5% at a stress of 150 MPa [205].

Fig. 3.25 shows displacement and resistance characteristics of a cantilever beam specimen made of a sputtered NiTiPd thin film [206]. A martensitic transformation with a hysteresis width of about 30 K is observed. Transformation temperatures M_f and A_f are 333 and 405 K, respectively. These values depend on both the chemical composition of the thin films and the temperature of the annealing process, as crystallization results in the formation of Ti_2Pd precipitates which reduce the effective Ti content of the matrix and may modify the intrinsic stress fields under certain conditions [207, 208]. NiTiPd thin films are used as a starting material for the development of microvalves for high temperature applications, see Sect. 5.3.3.4.

3.5 Development Trends

Besides exhibiting outstanding shape memory properties, NiTi alloys have many additional features that are attractive for applications such as high strength, good deformability, high corrosion resistance, and good biocompatibility. Therefore, a

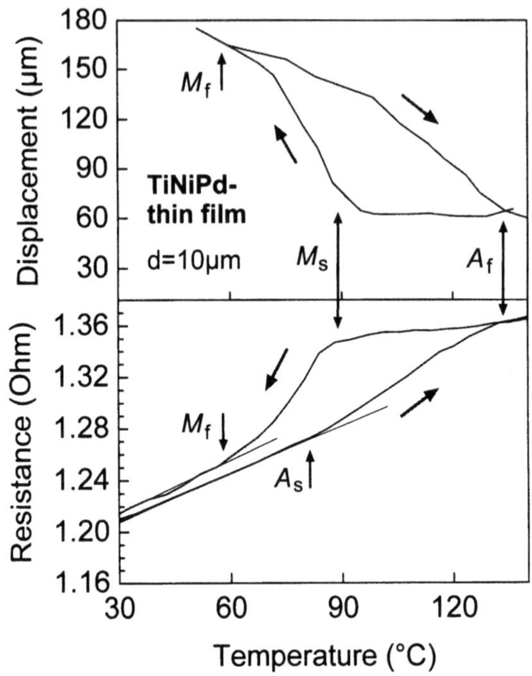

Fig. 3.25. Displacement and electrical resistance of a cantilever beam specimen made of a sputtered NiTiPd thin film (integral chemical composition: $Ni_{20}Ti_{52}Pd_{28}$, sputtering power: 300 W, annealing treatment: 843 K – 2.7 ks) [206]

number of fabrication processes have been developed to provide semi-finished products suitable for applications, e.g., rolling, drawing, sputtering, melt spinning, and sintering. Other SMA materials, for instance, those based on Cu or Fe, have a lower application potential from today's point of view.

Base materials for fabrication of microactuators are, in particular, foils and thin films in order to apply the planar technologies approved in microelectronics. Due to this reason, large efforts are undertaken worldwide to produce such materials with defined mechanical and thermal properties. In this context, important objectives for research and development are:

- Adjustment of transformation temperatures: The application range of SMA materials is limited by the available transformation temperatures. To further increase the application potential, materials with high transformation temperatures have to be to developed. Promising approaches include thin films of NiTiPd or NiTiHf.
- Adjustment of transformation hysteresis: For certain applications, it is desirable to adjust the characteristic of hysteresis. This can be achieved to a limited extent by controlling the chemical composition and conditions of heat treatment. Current research activities are concerned with the development of thin

film composites for the adjustment of transformation hysteresis. In this case, the stress condition in the SMA thin film is controlled via the thermal expansion coefficient of a substrate layer. For thin film composites consisting of NiTi-based alloys, it has been demonstrated that a hysteresis-free transformation can be achieved [208,209].

- Development of suitable sacrificial layers: Sacrificial layers are required for developing movable or transferable structures that remain stable during heat treatment of the SMA layers and remain etchant-resistant during chemical micromachining of the SMA layers.
- SMA layers with enhanced transducing properties: The shape memory properties of materials based on NiTi mainly depend on the temperature and stress. By allowing a larger variety of chemical compositions, SMA layers with additional degrees of freedom may be realized showing, e.g., an additional magnetic- or electric-field dependence. A recent example are ferromagnetic SMAs, which exhibit shape memory as well as magnetic properties. Presently, these materials are being explored intensively by several research groups, see, e.g., [210].
- Development of intelligent layer systems: By combining layers with various functionality (e.g., actuation, sensing, isolation, etc.), smart systems of compact size may be realized.

Currently, one major trend in microsystems technology is the development of systems with three-dimensional geometry. Thus, the development of SMA layer systems may only represent an intermediate stadium until suitable technologies become available, which may supersede planar technologies.

4 Shape Memory Actuation

Whereas design reveals talent, implementation reveals art.
–Marie Freifrau von Ebner-Eschenbach

Actuator development requires appropriate simulation tools and optimization methods as well as a systematic actuator design. Furthermore, technologies of sufficient precision and flexibility are needed to manufacture SMA actuators. Conventional NiTi actuators mostly consist of semi-products such as wires or plates, due to the bad material-cutting properties. Through the use of microtechnologies, however, microactuators with much higher geometrical complexity can be produced. The transition to SMA microactuation, therefore, results in a modified approach to actuator development. First, the aspects of simulation and optimization will be discussed. Then, the design of SMA microactuators and technological aspects will be addressed.

4.1 Overview

SMA actuation is based on the one-way effect. In principle, this effect can be used for generating motion and/or force. The heating power required is supplied either indirectly by variation of ambient temperature or directly by electrical heating of the SMA component. Thermal actuators react to temperature changes in the environment and, thus, may be operated both as temperature sensors and actuators. Electrically driven actuators carry out specific functions at ambient temperatures below the M_s temperature, which can be controlled via heating power. Further functions of SMA components may be structural or reset functions. This multifunctionality allows for intelligent solutions with extremely compact designs.

As shown in Fig. 4.1, the functions of SMA actuators are usually divided into three categories depending on the load case [190]:

(a) Free recovery: In the absence of an external load, a deformed SMA component returns into its original shape when heated above its A_f temperature. After this, further temperature changes do not cause any macroscopic shape changes.

(b) Constrained recovery: If a deformed SMA component is prevented from changing its shape when heated above its A_f temperature, a large force is exerted. This principle is applied successfully in fastener rings or electric connectors.

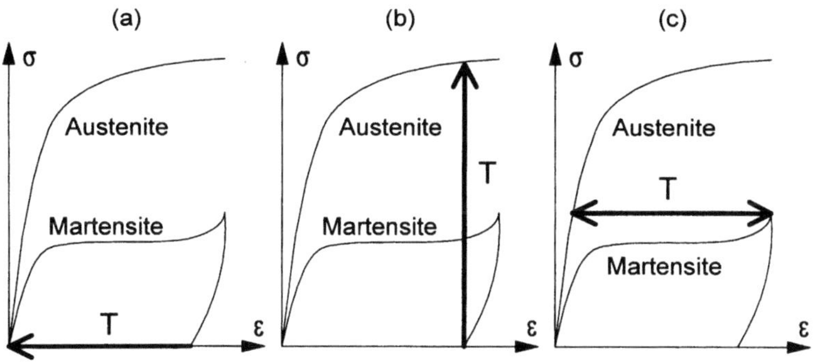

Fig. 4.1. Functional mechanisms of SMA actuators in the stress-strain diagram: (a) free recovery, (b) constrained recovery, (c) work production at constant reset force [190]

(c) Work production: If the SMA component overcomes an external reset force when heated above its A_f temperature, it generates motion and produces work. When cooled down, the reset force causes a movement back into the initial position. Thus, repeated heating and cooling produces periodic actuation cycles.

Fig. 4.1(c) shows reset movement for a constant reset force, whereby the stress remains unchanged. Depending on the choice of reset mechanism, other trajectories may be obtained in stress-strain space. The principle of work production with an external reset mechanism is applied in most SMA actuators. Exceptions are SMA actuators based on the two-way effect. However, it is difficult to adjust a defined two-way behavior in microstructures.

Fig. 4.2 shows various reset mechanisms using an SMA tension wire as an example. As shown in case (a), a weight produces a constant force level. Another widely used variant of the reset mechanism is that of a passive biasing spring (b). The use of a second SMA tension wire with opposite actuation direction leads to an antagonistic reset mechanism (c). Here, the antagonists are actuated alternately. Without actuation, the antagonists are in martensitic condition and act as a soft biasing element that can be deformed easily without requiring much work. Therefore, the antagonistic reset mechanism allows for larger displacements compared to biasing spring mechanisms. The case of decreasing reset force for increasing deflection is referred to as reverse biasing (d). Examples are the use of a magnetic actuator with opposite force direction or the use of a transmission increasing with displacement.

As the reset mechanism determines the stress profile in the SMA component, the hysteresis characteristic in the regime of phase transformation can be adjusted by selecting an appropriate reset mechanism.

Under pure tension or compression load, high forces but relatively small displacements can be reached. Larger displacements are achieved under torsion or bending loads. In these cases, however, the SMA material is loaded rather heterogeneously, such that only a fraction of it is used for work production.

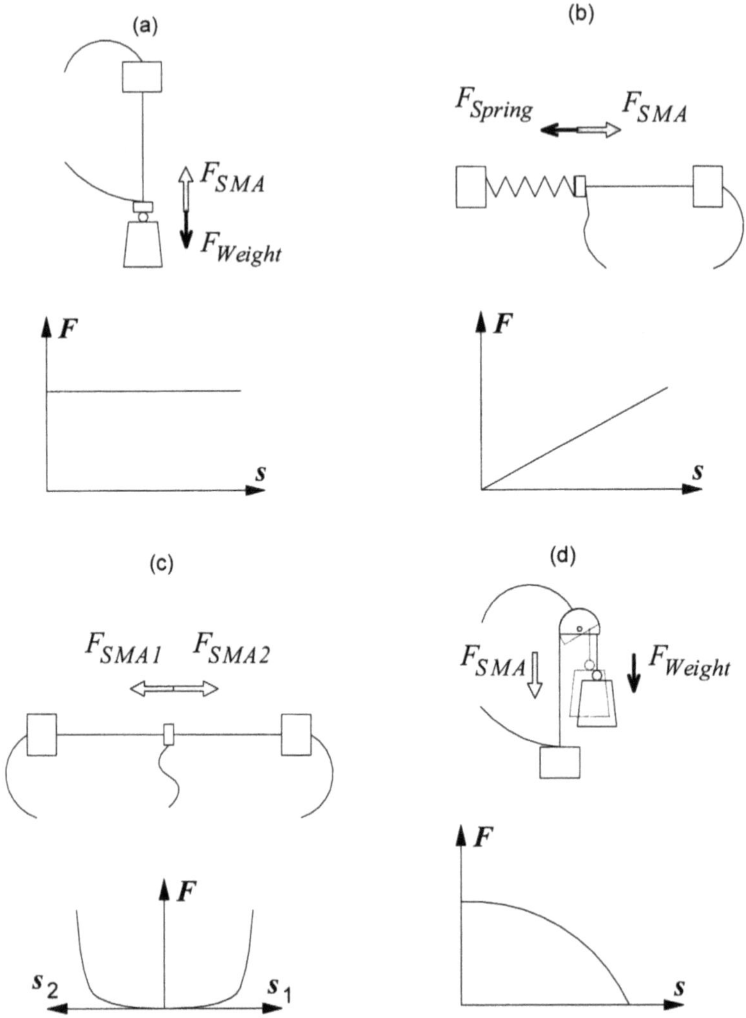

Fig. 4.2. Recovery mechanisms and corresponding force-displacement characteristics for the example of an SMA tension wire

Table 4.1 presents a survey of maximum achievable work densities and efficiencies of NiTi wires [8]. The overall low efficiencies result from the thermodynamically inefficient heat transfer during heating and cooling.

Frequently the physical behavior of SMA microactuators can no longer be estimated by simple analytical methods:

Table 4.1. Maximum achievable work densities, normalized to the actuator mass, and efficiencies of NiTi wires under different loads [8]

Load	Work density (J/kg)	Efficiency (%)
Tension	460	1.3
Torsion	80	0.23
Bending	4.6	0.013

- As the material can be micromachined laterally in nearly any way, a much larger variety of shapes can be produced in microsystems technology compared to conventional SMA technology.
- Generally, microactuators consist of closely coupled functional units that cannot be studied separately.
- SMAs exhibit a strongly nonlinear material behavior, pronounced hystereses, and fatigue above certain load limits.

For these reasons, numerical methods in the framework of field simulation are required for the design of SMA microactuators. Methods of three-dimensional field simulation and coupled multilevel simulation of SMA microactuators will be presented below.

4.2 Simulation

Microsystems applications demand simulation of mechanical, electrical, or thermal fields in components of arbitrary geometry. The most common method for field simulation is the finite element method (FEM) [211, 212]. The starting point is a physical problem in geometric space, which is subdivided into a finite number of elements. For each element, unknown variables like stress, temperature, electrical field, etc. are approximated by known functions depending on geometric points (knots) for the determination of the elements. Through the application of extremal principles or the method of Galerkin, a linear equation system is obtained for determining the unknown variables at the knots. Thereby, numerical integration over each element and summation over the space region of interest are performed. The linear equation system can be solved by methods of linear algebra.

As described in Sect. 3.3.2, various models have been developed for simulation of shape memory behavior. So far, however, only a few models have been implemented in a FEM program [168, 213–216]. The models presented in [213, 216] and [168] are only suitable for one-dimensional and two-dimensional problems, respectively. Though the models considered in [214, 215] allow three-dimensional load cases, they are restricted to pseudoelastic material behavior. Currently, there is a lack of tools for coupled mechanical, electrical, and thermal FEM simulation of SMA actuators allowing three-dimensional load cases. Therefore, a research program has been started at Forschungszentrum Karlsruhe for development of a three-dimensional FEM model for SMA microactuators [175, 217], as well as tools for coupled multilevel simulation [218].

4.2.1 Three-Dimensional Mechanical Model for SMA Actuators

Starting from Eq. 3.16 (Sect. 3.3.2.4), the problem of modeling the nonlinear behavior of SMA materials in martensitic condition can be attributed to finding a suitable expression for the transformation strain $\{\varepsilon\}^{tr}$ as a function of total strain $\{\varepsilon\}$. One option is to proceed as for conventional plastic materials, whereby isotropic material behavior is assumed.

Fig. 4.3 shows typical characteristics of effective stress $\bar{\sigma}$ as a function of effective strain $\bar{\varepsilon}$ and effective transformation strain $\bar{\varepsilon}^{tr}$. The effective stress $\bar{\sigma}$ is given by its Cartesian stress components:

$$\bar{\sigma} = \frac{1}{\sqrt{2}}[(\sigma_x - \sigma_y)^2 + (\sigma_y - \sigma_z)^2 + (\sigma_z - \sigma_x)^2 + 6(\tau_{xy}^2 + \tau_{yz}^2 + \tau_{zx}^2)]^{1/2} . \quad (4.1)$$

According to the von Mises criterion, quasi-plastic behavior begins under any state of stress when the effective stress $\bar{\sigma}$ exceeds a certain effective yield stress $\bar{\sigma}_0$. In the case of unloading, isotropic hardening is assumed. Thus, quasi-plastic behavior resumes only when the effective stress exceeds its previous maximum value independent of the subsequent state of stress.

As illustrated in Fig. 4.3, the contributions of elastic strain and transformation strain are characterized by the elastic modulus E and the slope $H' = \mathrm{d}\bar{\sigma}/\mathrm{d}\bar{\varepsilon}^{tr}$, respectively.

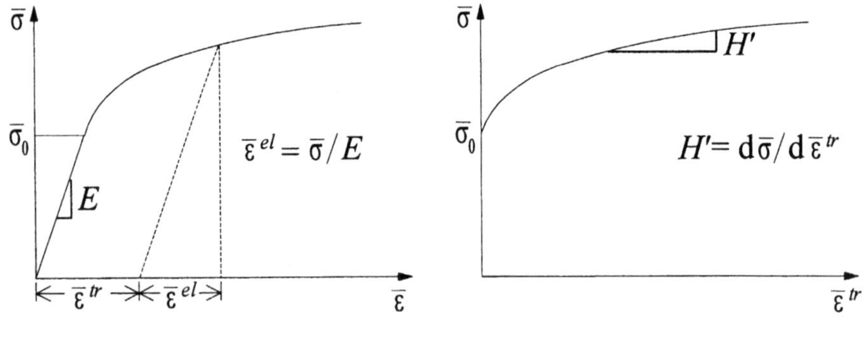

Fig. 4.3. Typical characteristics of effective stress and effective strain

Thus, the change of effective strain in martensitic condition can be expressed by:

$$\mathrm{d}\bar{\varepsilon} = \frac{\mathrm{d}\bar{\sigma}}{E^M} + \frac{\mathrm{d}\bar{\sigma}}{H'}, \tag{4.2}$$

where E^M is the elastic modulus in martensitic condition. The change of overall strain $\{\mathrm{d}\varepsilon\}$ is given by the changes of the elastic and quasi-plastic contributions $\{\mathrm{d}\varepsilon\}^{el}$ and $\{\mathrm{d}\varepsilon\}^{tr}$, respectively. From Eqs. 3.11, 3.13, and 3.15 the following expression is obtained:

$$\{\mathrm{d}\sigma\}^M = [E^M](\{\mathrm{d}\varepsilon\} - \{\mathrm{d}\varepsilon\}^{tr}). \tag{4.3}$$

The relationship between the changes of effective stress $\mathrm{d}\bar{\sigma}$ and the stress $\{\mathrm{d}\sigma\}$ can be determined by differentiation of both sides of Eq. 4.1:

$$\mathrm{d}\bar{\sigma} = \left\{ \frac{3\sigma'_x}{2\bar{\sigma}}\ \frac{3\sigma'_y}{2\bar{\sigma}}\ \frac{3\sigma'_z}{2\bar{\sigma}}\ \frac{3\tau_{xy}}{\bar{\sigma}}\ \frac{3\tau_{yz}}{\bar{\sigma}}\ \frac{3\tau_{zx}}{\bar{\sigma}} \right\}^T \left\{ \mathrm{d}\sigma_x\ \mathrm{d}\sigma_y\ \mathrm{d}\sigma_z\ \mathrm{d}\tau_{xy}\ \mathrm{d}\tau_{yz}\ \mathrm{d}\tau_{zx} \right\}$$

$$= \left\{ \frac{\partial\bar{\sigma}}{\partial\sigma} \right\}^T \{\mathrm{d}\sigma\}, \tag{4.4}$$

whereby the deviatoric stresses σ'_i follow from the Cartesian stress components according to $\sigma'_i = \sigma_i - 1/3\sum_j \sigma_j$.

The relationship between the changes of effective transformation strain $\mathrm{d}\bar{\varepsilon}^{tr}$ and transformation strain $\{\mathrm{d}\varepsilon\}^{tr}$ can be determined in an analogous way. As a result, the Prandtl-Reuss relations are obtained:

$$\{\mathrm{d}\varepsilon\}^{tr} = \left\{ \frac{\partial\bar{\sigma}}{\partial\sigma} \right\} \mathrm{d}\bar{\varepsilon}^{tr}. \tag{4.5}$$

Based on the Prandtl-Reuss relations 4.5 and Eqs. 4.3 and 4.4, the following relationship between the changes of effective transformation strain $\mathrm{d}\bar{\varepsilon}^{tr}$ and the total strain $\{\mathrm{d}\varepsilon\}$ results:

$$\mathrm{d}\bar{\varepsilon}^{tr} = \frac{\left\{ \frac{\partial\bar{\sigma}}{\partial\sigma} \right\}^T [E^M]}{H' + \left\{ \frac{\partial\bar{\sigma}}{\partial\sigma} \right\}^T [E^M] \left\{ \frac{\partial\bar{\sigma}}{\partial\sigma} \right\}} \{\mathrm{d}\varepsilon\} = [W]\{\mathrm{d}\varepsilon\}. \tag{4.6}$$

Finally, the relationship between the changes of stress and total strain in martensitic condition is obtained via the Prandtl-Reuss relations 4.5 and Eqs. 4.3 and 4.6, which can be written in the form of Hooke's law:

$$\{\mathrm{d}\sigma\}^M = \left([E^M] - [E^M] \left\{ \frac{\partial \overline{\sigma}}{\partial \sigma} \right\} [W] \right) \{\mathrm{d}\varepsilon\} = [E^{M*}] \{\mathrm{d}\varepsilon\}. \tag{4.7}$$

In this expression, thermal expansion is not included, but may be taken into account by a further additive term given by Eq. 3.17. Eqs. 4.6 and 4.7 are particularly valid for the case of perfect plasticity, as no divergence occurs in the case of $H'=0$.

The relationship between the changes of stress and total strain in austenitic condition is given by Hooke's law, as for conventional elastic materials, see Eq. 3.12.

The change of total stress $\{\mathrm{d}\sigma\}$ in the regime of phase transformation is given by the changes of the martensitic and austenitic phase fractions according to the rule of mixture. Differentiation of Eq. 3.10 results in:

$$\{\mathrm{d}\sigma\} = \xi \{\mathrm{d}\sigma\}^M + (1-\xi)\{\mathrm{d}\sigma\}^A + \mathrm{d}\xi(\{\sigma\}^M - \{\sigma\}^A). \tag{4.8}$$

4.2.2 FEM Formulation of Mechanical Model for SMA Actuators

Assuming that the dynamics of an SMA component is determined by heat transfer, inertia and viscous damping can be neglected. The static behavior of a three-dimensional body follows from the force balance between internal and external forces [212]:

$$\int [B]^T \{\sigma\} \mathrm{d}V = \{P\}, \tag{4.9}$$

where matrix $[B]$ fulfills the strain-displacement relationship $\{\varepsilon\}=[B]\{U\}$. In the case of geometrical nonlinearity, matrix $[B]$ depends on the displacements. Then, the expression $[B]=[B_0]+[B_l(U)]$ given in Ref. [219] is used. The term $\{P\}$ denotes the matrix of external forces, which in general depends on the unknown displacements of knots $\{U\}$. Due to the nonlinear behavior of SMAs in martensitic condition, Eq. 4.9 is nonlinear even for small displacements.

Therefore, FEM analysis is performed by using an incremental method. An approved procedure is the tangential stiffness method, which is illustrated in Fig. 4.4 for the one-dimensional case. The development of quasi-plastic displacements of knots is traced by applying small load steps. Assuming that under loads $\{P\}^{(n-1)}$ the correct displacements $\{U\}^{(n-1)}$ are known, the displacements caused by the next load step $\{P\}^{(n)}$ can be computed from the linear equation system:

$$\int [B]^T \{\Delta\sigma\}_i^{(n)} \mathrm{d}V = \{\Delta P\}_i^{(n)}. \tag{4.10}$$

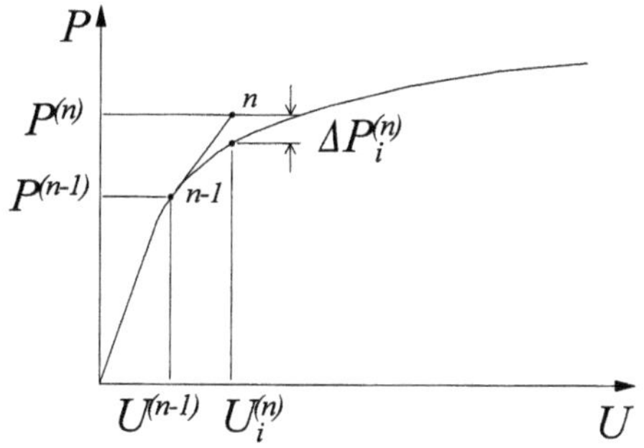

Fig. 4.4. Typical load step of the tangential stiffness method at a point in load-displacement diagram (one-dimensional representation)

The term $\{\Delta P\}_i^{(n)}$ denotes the non-balanced forces at the *i*th iteration, which is given by the difference between inner forces induced by the stress in the elements and the external load $\{P\}^{(n)}$:

$$\{\Delta P\}_i^{(n)} = \int [B]^T \{\sigma\}_i^{(n)} \mathrm{d}V - \{P\}^{(n)}. \tag{4.11}$$

According to Eq. 4.8, the change of total stress $\{\Delta\sigma\}_i^{(n)}$ results from the changes of stress in martensitic (Eq. 4.7) and austenitic conditions (Eq. 3.12). For each iteration a constant phase fraction $\xi_i^{(n)}$ can be assumed. By inserting $\{\Delta\sigma\}_i^{(n)}$ in Eq. 4.10 and making use of the strain-displacement relationship, we obtain:

$$[K]_i^{(n)} \{\Delta U\}_i^{(n)} = \{\Delta P\}_i^{(n)}, \tag{4.12}$$

where $[K]_i^{(n)}$ is the tangential stiffness matrix:

$$[K]_i^{(n)} = \int [B]^T \xi [E^{M*}]_i^{(n)} [B] \mathrm{d}V + \int [B]^T (1-\xi) [E^A]_i^{(n)} [B] \mathrm{d}V. \tag{4.13}$$

Using Eq. 4.12, an iterative improvement $\{\Delta U\}_i^{(n)} = \{U\}_{i+1}^{(n)} - \{U\}_i^{(n)}$ of the previous approximation $\{U\}_i^{(n)}$ is obtained. The volume integrals are separated into a sum of integrals over the volumes of individual elements, which are calculated later on for each element at certain sampling points (Gauss points). Then, the contributions of each element are assembled again to produce the tangential stiffness matrix of the overall volume. The resulting equation system can now be solved.

The sequence of FEM simulation of the mechanical behavior of SMA actuators is illustrated in Fig. 4.5.

Calculation of the $(i+1)$th iteration step is performed in the following way:

(1) Starting with the known expressions for $[K]_i^{(n)}$ and $\{\Delta P\}_i^{(n)}$, the overall displacements $\{\Delta U\}_i^{(n)}$ are calculated via Eq. 4.12. From the overall displacements the element displacements $\{\Delta u\}_i^{(n)}$ are determined.

(2) For each sampling point in the elements, the change of overall strain $\{\Delta \varepsilon\}_i^{(n)} = [B]\{\Delta u\}_i^{(n)}$ is determined. The change of overall strain $\{\Delta \varepsilon\}_i^{(n)}$ contains the contributions of elastic and transformation strain in martensitic and austenitic conditions.

(3) Based on the change of overall strain $\{\Delta \varepsilon\}_i^{(n)}$, the changes of effective transformation strain $\Delta \bar{\varepsilon}_i^{tr,(n)}$ (Eq. 4.6), of transformation strain $\{\Delta \varepsilon\}_i^{tr,(n)}$

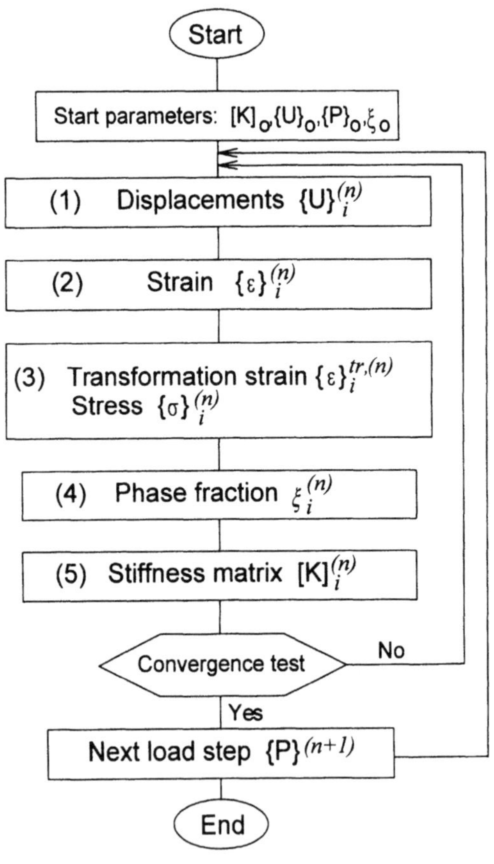

Fig. 4.5. Flow chart for FEM simulation of the mechanical behavior of SMA actuators

(Eq. 4.5), of effective stress $\Delta\bar{\sigma}_i^{M,(n)}$ ($\Delta\bar{\sigma}_i^{M,(n)} = H_i'^{(n)}\Delta\bar{\varepsilon}_i^{tr,(n)}$), and of the stress in martensitic condition $\{\Delta\sigma\}_i^{M,(n)}$ (Eq. 4.3) can be calculated successively for each sampling point. Furthermore, the change of stress in austenitic condition $\{\Delta\sigma\}_i^{A,(n)}$ (Eq. 3.12) can be determined. The calculated changes are used to update the corresponding quantities, e.g., $\bar{\varepsilon}_{i+1}^{tr,(n)} = \bar{\varepsilon}_i^{tr,(n)} + \Delta\bar{\varepsilon}_i^{tr,(n)}$. From the actual phase fraction $\xi_i^{(n)}$ the overall stress $\{\sigma\}_{i+1}^{(n)}$ (Eq. 3.10) and, hence, the effective overall stress $\bar{\sigma}_{i+1}^{(n)}$ (Eq. 4.1) are determined.

(4) Now, the phase fraction is updated to $\xi_{i+1}^{(n)}$ using Eq. 3.6 or 3.7 depending on the direction of phase transformation.

(5) After calculating the slope $H_{i+1}'^{(n)} = \Delta\bar{\sigma}_{i+1}^{M,(n)} / \Delta\bar{\varepsilon}_{i+1}^{tr,(n)}$ and the expressions $\{\partial\bar{\sigma}/\partial\sigma\}_{i+1}^{(n)}$ (Eq. 4.4), $[E^{M*}]_{i+1}^{(n)}$ (Eq. 4.7) and $[E^{A}]_{i+1}^{(n)}$ (Eq. 3.12), the new tangential stiffness matrix for each element $[\mathrm{k}]_{i+1}^{(n)}$ can be determined by Eq. 4.13. Then, the tangential stiffness matrix of the overall volume $[K]_{i+1}^{(n)}$ is produced from the contributions of each element $[\mathrm{k}]_{i+1}^{(n)}$ and, finally, the load for the next iteration step $\{\Delta P\}_{i+1}^{(n)}$ (Eq. 4.11) is determined.

Steps 1–5 have to be repeated until the solution converges, i.e., once a norm of the non-balanced forces $\{\Delta P\}_i^{(n)}$ is smaller than a given tolerance. After convergence is reached, the subsequent load step $\{P\}^{(n+1)}$ can be computed. In order to improve convergence, particularly in the case of geometrical nonlinearity, the mechanical FEM analysis may be performed by applying a Newton Raphson algorithm [219]. In this case, however, the change of phase fraction $\mathrm{d}\xi$ needs to be calculated explicitly.

4.2.3 Simulation of Electrical Performance

The presented mechanical model is based on the assumption of a constant temperature. This ideal case is relevant to the fundamental understanding of SMA components. However, real SMA actuators are mostly driven directly or indirectly by an electrical current, which leads to local heat generation and, consequently, to a temporally and spatially varying temperature profile. Thus, for real cases, the electrically induced heating power $\dot{Q}$ must also be calculated. In the following, the case of direct electrical heating of an SMA actuator will be considered.

At first, the profile of electrical current j has to be determined in order to calculate $\dot{Q}$. The corresponding electrical potential fulfills the differential equation:

$$\{\rho^{-1}\}\Delta\Phi = 0 \tag{4.14}$$

and the Neumann boundary condition at the electrical connections:

$$\{\rho^{-1}\}\vec{\nabla}\Phi = \{j_0\}\,. \tag{4.15}$$

The term $\{\rho^{-1}\}$ is the matrix of electrical conductivity. Assuming a temperature-independent electrical conductivity, Eqs. 4.14 and 4.15 can be solved, as for conventional materials, by standard FEM analysis, see, e.g., [211]. From the profile of electrical potential the current density and thus the heating power are determined:

$$\dot{Q} = \int (\{\rho\}\{j\})^T \{j\}\mathrm{d}V\,. \tag{4.16}$$

In a more accurate treatment, the nonlinear change and hysteresis of the electrical conductivity may be simulated according to the mechanical treatment by taking the phase fraction–dependent contributions of martensitic and austenitic conductivity into account:

$$\{\mathrm{d}\rho^{-1}\} = \xi\{\mathrm{d}\rho^{-1}\}^M + (1-\xi)\{\mathrm{d}\rho^{-1}\}^A + \mathrm{d}\xi(\{\rho^{-1}\}^M - \{\rho^{-1}\}^A)\,. \tag{4.17}$$

Phase fraction ξ and its derivative $\mathrm{d}\xi$ follow from the kinetic Eqs. 3.6 and 3.7.

4.2.4 Simulation of Thermal Performance

For determination of temporally and spatially varying temperature profiles in SMA actuators, the heat exchange with the environment has to be taken into account, in addition to the electrically induced heating power $\dot{Q}$. Below 200 °C, the heat exchange is mainly governed by heat conduction and convection. In this case, temperature profiles can be determined by solving differential equations 3.18 and 3.19 by a thermal FEM analysis, see, e.g., [211]. At higher temperatures, heat radiation must also be considered.

In contrast to conventional materials, the latent heat of phase transformation has to be taken into account for thermal simulation of SMA materials, which is released upon cooling and consumed upon heating. For constant stress, the latent heat is proportional to the thermal change of martensitic phase fraction:

$$h(T) = \Delta H \frac{\mathrm{d}\xi}{\mathrm{d}T} \tag{4.18}$$

and thus can be determined by Eqs. 3.6 and 3.7. The term ΔH is the enthalpy of phase transformation, which can be determined experimentally, e.g., by heat calorimetry.

4.2.5 Coupled Multilevel Simulation

The differential equations for calculating the mechanical, electrical, and thermal performance of real SMA actuators are coupled with each other. Thus, a self-consistent solution is desired, which can be accomplished by coupling the different FEM programs used for solving each physical layer. For this purpose, special program routines are required. Special coupling programs provide the exchange of intermediate results between the different FEM programs as well as the update of boundary conditions, material, and geometry data. A superordinated control program is installed that launches the FEM programs according to a predefined sequence listed in its input file [218]. The execution sequence needs to be repeated iteratively to obtain self-consistent results. For time-dependent calculations, an individual time is attributed to each sequence.

The simulation procedure for self-consistent calculation of the coupled mechanical, electrical, and thermal performance of an SMA actuator is illustrated in Fig. 4.6 [217].

In this case, the simulation is performed in five steps:

(1) Provision of simulation sequence: Orders, times, and control commands are entered in an input file for the different programs (FEM programs, coupling programs) and transferred to the control program. The control program calls the different programs and passes times, file names, and control commands correspondingly.

(2) Geometry processing: The initial geometry of an SMA actuator is specified and a mesh is generated for the following FEM program.

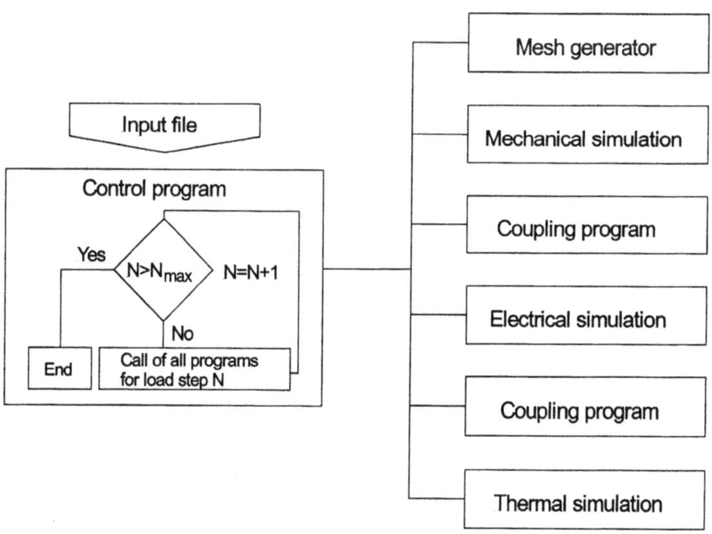

Fig. 4.6. Scheme of coupled multilevel simulation of an SMA actuator

(3) Mechanical simulation: For a given load, strains and stresses in the SMA actuator are calculated. Later on, the FEM results are processed and boundary conditions are set by a coupling program for the subsequent FEM simulation. If necessary, another cycle through the mesh generator is performed.

(4) Electrical simulation: For a given electrical potential or a given current density specified at the boundaries, the profile of current density in the SMA actuator is calculated. Then a coupling program is executed, which determines the corresponding profile of heating power and passes the results to the subsequent FEM program. Furthermore, boundary conditions are set for the subsequent FEM simulation.

(5) Thermal simulation: The temperature profile is calculated as a function of time based on the locally induced heating power and given boundary conditions for the heat exchange of the SMA actuator with its environment.

The presented models and procedures allow quantitative predictions on the mechanical, electrical, and thermal performance, e.g., of SMA microvalves and SMA linear actuators, as will be discussed in Chaps. 5 and 6, respectively.

4.3 Design Optimization

The usability of the shape memory effect is strongly dependent on the component geometry, as can already be demonstrated for simple SMA components. Fig. 4.7 shows, for instance, the von Mises stress profile on the surface of an SMA double-beam element loaded vertically to beam direction.

In the black regions, no stress exists, whereas in the white regions, stress is maximum. It is evident that maximum stress values occur only locally near the beam suspension. Therefore, it is concluded that the shape memory effect is used only in a very small volume fraction of the component. Work density, accordingly, is small. Furthermore, the regions with accumulated stress are very sensitive in terms of mechanical overload and fatigue. These cohesions are generally true for the case of more complex SMA microcomponents with arbitrary lateral shape. Hence, the development of SMA microactuators demands for optimization of profile and level of stress.

The optimization criteria with regard to the mechanical properties of SMA components may be summarized as follows:

(O1) For a given load, a homogeneous stress profile should be generated.

(O2) Maximum stress should be high to reach a sufficiently high work density.

(O3) A material-dependent stress limit should not be exceeded to prevent mechanical overload and limit fatigue effects. For NiTi, for instance, the values given in Table 3.4 (Sect. 3.4.3.4) apply.

Additional criteria must be found to optimize the thermal properties of SMA components. For example, Fig. 4.8 shows the temperature profile generated by an electrical current in an SMA double-beam element.

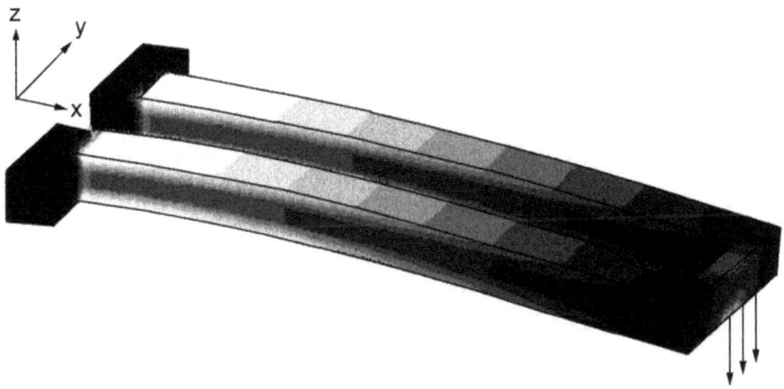

Fig. 4.7. Calculated von Mises stress profile in an SMA double-beam element. Black regions: no stress, white regions: maximum stress. The arrows indicate loading at the end of the beam

Due to heat conduction, regions near the electrical contacts are better cooled than regions farther away. Moreover, convection causes additional cooling, which increases with increasing surface-to-volume ratio.

In case the beam element shown in Fig. 4.8 is heated electrically, phase transformation first takes place at the front end of the beam. With increasing heating, the phase transformation boundary moves toward the electrical contacts. For complete phase transformation in the coolest regions near the electrical contacts, a relatively high current is required. This may already lead to a temperature overload at the front end of the beam.

The example of the SMA double-beam element demonstrates that regions of maximum stress, in general, are not identical to regions of maximum temperature. As a consequence, a small efficiency may be achieved, since a high heating power is needed for thermal activation of regions of high stress.

Fig. 4.8. Calculated temperature profile in an SMA double-beam element indicated by different gray levels. Maximum temperature is indicated in black

Further thermal optimization, therefore, requires the following additional criteria to be fulfilled:

(O4) A given current distribution should generate a temperature profile as homogeneous as possible.

(O5) Areas of maximum temperature should be identical with areas of maximum stress.

(O6) A certain material-dependent temperature limit should not be exceeded to prevent thermal overload and the associated change or loss of the shape memory effect. For NiTi, e.g., changes of the shape memory effect due to diffusion processes are to be expected above a temperature of about 250 °C.

In general, thermal, mechanical, and other optimization criteria, if applicable, cannot be fulfilled at the same time. Hence, application-specific compromise solutions have to be found for the design of SMA actuators. Below, options for mechanical and thermal optimization will be presented.

4.3.1 Mechanical Design Optimization

Depending on the complexity of the component geometry, stress profiles in SMA components can be optimized either analytically or by iterative FEM simulation. Simple analytical solutions are found, e.g., in the case of an SMA beam element. Assuming a rectangular beam of constant thickness h, which is oriented in x-direction, the stress along the upper or lower surface $\sigma_S(x)$ is given by the distance between the neutral fiber and the upper or lower surface $h/2$, the bending moment $M_b(x)$, and the axial surface moment of second degree $I_y(x)$:

$$\sigma_S(x) = \frac{M_b(x)}{I_y(x)} \frac{h}{2} . \tag{4.19}$$

Mechanically optimized beam geometries should fulfill the criteria (O1) through (O3). This means that the stress along the beam surface $\sigma_S(x)$ should be constant and, in addition, near a material-dependent stress limit σ_{limit}. Depending on the application, the stress limit σ_{limit} should be selected such that a sufficient number of load cycles is possible. This can be summarized by the boundary condition:

$$\sigma_S(x) = \sigma_{limit} . \tag{4.20}$$

By inserting the boundary condition 4.20 into Eq. 4.19 and taking into account the axial surface moment for a rectangular beam geometry $I_y(x)=h^3w(x)/12$, the following expression for the beam width $w(x)$ is obtained:

$$w(x) = \frac{6M_b(x)}{\sigma_{limit} h^2}. \tag{4.21}$$

Eq. 4.21 gives the stress-optimized width profile of a beam element of constant thickness. In particular, a linear increase in width from the beam end to the point of suspension is obtained for a point load. More complicated surface loads lead to nonlinear width profiles [220].

A suitable numerical optimization technique is the so-called computer-aided optimization (CAO) method, which was originally developed at the Forschungszentrum Karlsruhe for weight optimization in mechanical engineering [221]. Since then, the original version of the CAO method has been made commercially available as a supplementary program for the packages ANSYS and ABAQUS. The underlying idea is to simulate adaptive biological growth, e.g., of trees or bones. Trees reduce local stress concentrations caused by external loads by a locally increased or reduced growth in the outer range of the trunk. Growth of bones is characterized, besides by the addition of material at strongly loaded points, by the reduction of material at points of small loads, which results in weight optimization. The CAO method reaches this objective by a combination of mechanical and thermal FEM simulations without the need to know any functional relationships.

At first, the local von Mises stress profile $\sigma_{mises}(x,y,z)$ is determined by mechanical FEM simulation. Then, the stress distribution $\sigma_{mises}(x,y,z)$ is converted into a fictitious temperature distribution $T(x,y,z)$. During subsequent thermal FEM simulation, it causes local swelling in the range of specially introduced swelling layers on the component surface. By this mathematical trick, the local von Mises stress distribution is converted into a change of geometry. If appropriate swelling layers and termination criteria of the iterative simulation procedure are selected, this change of geometry leads to a continuous reduction of stress maxima and, hence, to a minimization of the component load.

By modifying the CAO method, stress homogenization can be achieved instead of stress minimization and, consequently, the mechanical optimization criteria (O1) through (O3) can be fulfilled [193]. A homogeneous stress distribution near the stress limit σ_{limit} is reached by permitting growth only in those areas where the stress is above the given limit σ_{limit}. In regions where the stress drops below σ_{limit}, optimization is terminated. The procedure for stress optimization of SMA components is summarized in Fig. 4.9.

Optimization is performed in three steps:

(1) Geometry processing: The initial geometry of the SMA component is specified. Then, a mesh is generated for the subsequent FEM program.

(2) Mechanical simulation: The stress distribution is calculated for a characteristic load case. To adjust the stress near the given limit σ_{limit}, the von Mises stresses calculated for each element are compared with σ_{limit}. The subsequent coupling program is called up if the von Mises stress for an element exceeds σ_{limit}. By means of the coupling program, the calculated stress dis-

tribution is converted into a fictitious temperature profile. Moreover, the boundary conditions are specified for the subsequent simulation.

(3) Thermal simulation: The expansion of the swelling layer is calculated for the given fictitious temperature profile. Only those elements are considered whose von Mises stress is higher than the given limit σ_{limit}. Mechanical loads are not considered in this simulation step. Then, the calculated coordinate shifts are added to the node geometries from (2) and the mesh is updated.

Steps (2) and (3) have to be repeated several times, until a sufficiently homogeneous stress state develops near the stress limit σ_{limit}. When the von Mises stresses of all elements of the swelling layer are below σ_{limit}, optimization is complete.

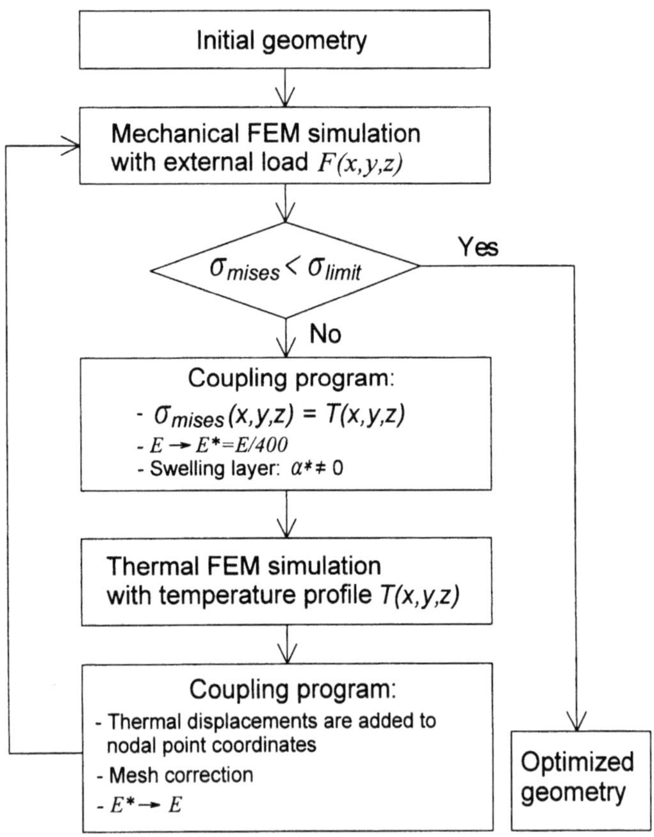

Fig. 4.9. Flow chart for stress optimization of SMA components using a modified CAO method. The parameters E, E*, and α* denote E-modulus for mechanical simulation, E-modulus and thermal expansion coefficient for thermal simulation, respectively

4.3.2 Thermal Design Optimization

Homogenization of the temperature profile by pure geometry optimization leads to a conflict with the mechanical optimization described above. This is evident already from the example on beam elements discussed above. For a constant beam width, stress decreases linearly in beam direction, while temperature increases as shown in Figs. 4.7 and 4.8, respectively. If heat conduction predominates, temperature increases even quadratically, as can be seen by integrating Eq. 3.18 for the stationary case ($\dot{T} = 0$):

$$T(x) = T_A + \frac{\dot{Q}}{2\lambda} x^2 . \qquad (4.22)$$

At the ends of the beam, the following boundary conditions have been considered:

$$T = T_A , \quad \text{at the beginning of the beam } x\text{=0 and}$$

$$-\lambda \frac{dT}{dx} = 0 , \quad \text{at the beam end } x\text{=}L. \qquad (4.23)$$

For optimization of temperature profile, it is, therefore, more reasonable to maintain the mechanically optimized geometry, if possible, and adjust the temperature profile by optimization of the arrangement of heat sources and sinks using methods of electrical engineering. This approach allows for a largely separate thermal optimization and helps to avoid bad compromise solutions in the development of optimized SMA microactuators. The effect of heat sources and sinks is analyzed by means of thermal FEM simulations.

4.4 Conventional SMA Actuators

Since the first publications on the shape memory effect, SMA components have entered the market only gradually. In the past years, applications on the basis of the pseudoelastic effect rose strongly, e.g., spectacle frames, mobile phone antennas, braces, stents. Furthermore, a number of commercially successful actuator products have been established to date that make use of the shape memory effect in SMA semi-products such as wires, helical springs or plates.

4.4.1 Overview

An overview of several major SMA actuators is given in Table 4.2. The oldest and so far most successful actuator applications of SMAs are tube and pipe couplings [222]. They are characterized by a simple assembly, small dimensions, high contact stresses (500–700 MPa), and a high reliability. These advantages are of particular importance in, e.g., aviation and space technology.

As the couplings have to remain in austenitic condition during operation, assembly takes place far below the operation temperature range. Presently, two assembly technologies are being applied. The low-temperature method uses liquid nitrogen for the storage and assembly of NiTi couplings. Recently, alloys based on NiTiNb with an extremely high hysteresis have been made available, where no cooling of the coupling is required. In this case, a short heat treatment is sufficient for assembly ("heat-to-shrink" technology).

Meanwhile, the mechanism of constrained recovery is also used commercially in a number of other applications. As an example, electrical connection systems should be mentioned. In spite of high tolerances, such connection systems allow a force-free insertion of the contact pins and provide for a good electrical contact after a short heat treatment [223]. In all applications of this type, thermal actuators with one-way effect are applied.

Moreover, thermal actuators are used for work production. In most cases, they are equipped with an SMA helical spring and a passive biasing spring. Typical examples are thermal valves, as shown schematically in Fig. 4.10. In this case, the SMA actuator works both as a sensor for temperature changes in the environment and as an actuator controlling a fluid flow. This concept results in a simple and compact setup. Due to the high work output available, high pressures and large fluid flows can be controlled. A widely used application are, e.g., anti-scald safety valves in shower heads [224].

So far, electrically driven SMA actuators have been used only in a few applications. This is mainly due to the relatively high heating powers caused by a small efficiency (see Table 4.1) and macroscopic dimensions. Numerous prototypes have been developed for applications in robotics. Here, the advantages, such as a high energy density, compact design, electrical heating, and noise-free operation, are of particular use. Figure 4.11 shows the example of a robot hand operated with SMA wires [225].

Table 4.2. Examples of conventional macroscopic SMA actuators

Shape memory effect	Functional mechanism	Application	Ref.
One-way effect	Thermal actuation, constrained recovery	Couplings	[222]
		Fastener rings	[226]
		Electrical connection systems	[223]
One-way effect	Thermal actuation, work production with passive biasing mechanism	Anti-scald safety valves	[224]
		Air conditioners	[227]
		Circuit breakers	[228]
		Temperature switches (toasters, coffee machines)	[229]
One-way effect	Electrical actuation, work production with passive biasing mechanism	Light controls	[230]
		Grippers	[225]

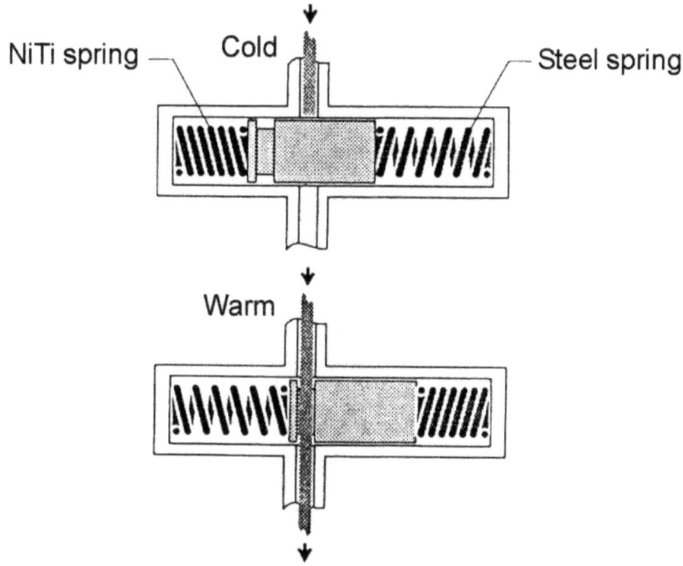

Fig. 4.10. Setup and operation principle of a thermal valve [230]

Fig. 4.11. Robot hand with SMA wire actuators [225]

A new development trend in conventional SMA actuator technology are composites of SMA components and a surrounding matrix, e.g., made of silicone, to obtain multifunctional properties. In this case, the SMA components may take over active movement functions, while the surrounding matrix fulfills the reset function and other functions if necessary, such as chemical safety functions, cooling, sensing, etc. Examples of such composite actuators are SMA endoscopes [231] and SMA grippers [232].

4.4.2 Design

Designing conventional actuators usually starts from SMA semi-products, which nowadays include pipes, cylinders, wires, helical springs, and plates. For predetermined set values, the required dimensions of the SMA semi-product are specified taking into account the load case and possibly needed reset elements. Geometry optimization is limited mainly to the optimization of the semi-product dimensions. When designing helical tension or compression springs, for instance, the calculation formulas for linear-elastic springs according to DIN 2089 can be applied. This leads to linear shear stress–shear characteristics for the austenitic and martensitic condition. In martensitic condition, the linear behavior represents only a rough approximation. When introducing an effective modulus for a given strain, however, this approximation is associated with a still tolerable error. The precise characteristics are usually determined empirically using test springs.

Under tension or compression load, helical springs are loaded inhomogeneously with the shear stress and the shear reaching maximum values on the inner side of the winding. Based on certain limit values for the shear stress τ_G and shear γ_G (typical values for NiTi are, e.g., τ_G=100 N/mm^2 and γ_G = 1.5%) and set values on the force and displacement, it is then possible to determine the parameters of wire diameter, mean winding diameter, and number of spring windings. The force-displacement characteristic of the biasing spring has to be added to the set values. When designing the biasing spring, the maximum stress must remain below the material-dependent yield stress. As illustrated in Fig. 4.12, force-displacement characteristics for SMA spring and biasing spring are obtained, which determine the resulting actuation stroke.

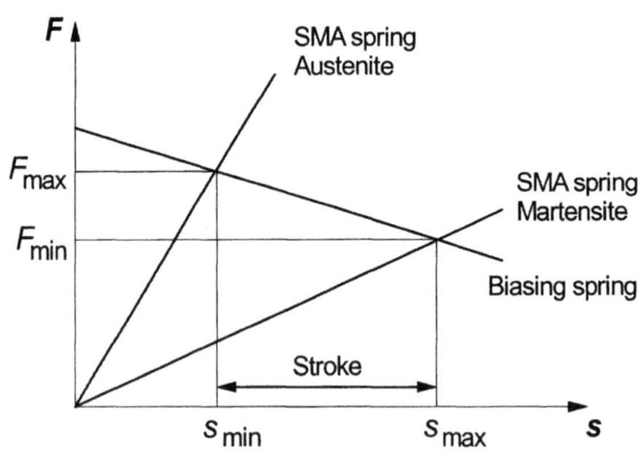

Fig. 4.12. Schematic of the force-displacement characteristics of an SMA spring in austenitic and martensitic conditions and a biasing spring. Stiffness of the biasing spring determines the maximum possible stroke of the SMA actuator. F_{min}, F_{max}, s_{min}, and s_{max} denote the minimum and maximum forces and displacements

At the operation temperatures of most SMA components, indirect heating and cooling occurs mainly by heat conduction and convection. Rapid heat transfer by heat conduction requires a direct thermal contact with the heat source or heat sink. For example, PTC resistors are very much suited for heating. When they reach a certain temperature, their electrical resistance is increased considerably, leading to a limitation of temperature. Good heat transfer by convection requires large transfer surfaces and temperature gradients.

For SMA helical springs, direct heating with electrical current is employed. In this case, the generated heating power is subject to Joule's law. The reaction time can simply be estimated from the ratio between heat energy Q to be applied, which consists mainly of the fractions of specific heat and latent heat of phase transformation, and the heating power RI^2. When neglecting cooling effects during heating (adiabatic case), a typical reaction time of 2 s with a power consumption of 950 mW results for a NiTi wire of 100 mm length and 0.3 mm diameter.

When designing conventional SMA actuators as outlined above, the design criteria (O1) through (O6) presented in Sect. 4.3 are considered only partly:

(O1)–(O3) By specifying a maximum stress load, critical stress limits are observed. Due to simple semi-product geometries, this can be done within the framework of analytical estimations. However, by limiting the geometry of conventional SMA components to given semi-product geometries, mechanical optimization options are strongly restricted. Except for simple SMA tension elements, an inhomogeneous stress profile generally prevails in the SMA component. As a consequence, work density and possibly the maximum number of actuation cycles are reduced.

(O4)–(O6) Presently, there is no commercially available simulation tool to calculate temperature profiles in electrically driven SMA components. For this reason, a thermal optimization can only be performed in an approximative way. In many applications, a more or less inhomogeneous temperature distribution exists over the cross section and length of the SMA components, which adversely affects the reaction time and possibly the service life.

4.4.3 Fabrication Technologies

The fabrication of SMA semi-products is described in literature in detail [233]. The most commonly used technology for adjusting the composition of NiTi alloys is based on the compression of Ni resin and Ti sponge to disc-shaped moldings that are welded to a cylindrical ingot. In this process, the chemical composition must be observed precisely, as the transformation temperatures are sometimes extremely dependent on the stoichiometry. To keep the temperature difference of Ni-rich NiTi below 2 K, e.g., the Ni content must be controlled with an accuracy of at least 0.05 at.%. Then, the ingot is remolten several times to reach a composition as homogeneous as possible. Due to the high reactivity of Ti, high requirements with respect to the purity of the chemical composition must be complied with. Subsequently, blanks are produced by extrusion. These blanks are used to produce semi-products, e.g., by rolling and drawing.

By means of cold rolling, the blank thickness is reduced by 20% at maximum in order to limit the number of introduced defects. Subsequently, these defects have to be healed by a heat treatment. Consequently, strong thickness reductions require the cold rolling and heat treatment processes to be repeated several times.

Due to the special mechanical properties of SMA materials, three-dimensional formation by conventional mechanical processing methods (e.g., milling) is rather difficult. This is a main reason why conventional SMA actuators are nearly exclusively made up of SMA semi-products. The high effort already required for the manufacture of SMA semi-products leads to relatively high material costs.

Using a thermomechanical pretreatment, defects can be influenced specifically. Usually, rolled SMA semi-products are available in work-hardened form, as a result of which the shape memory effect is strongly impeded. In this case, a single heat treatment (conditioning) is needed to reduce the defect density. By this process, transformation temperatures can be adjusted to a certain degree. At the same time, the memory shape for the one-way effect is specified by the shape given during conditioning. Subsequently, the two-way effect can be adjusted by certain training methods [153].

Heat treatment is adversely affected by oxide layers, which, e.g., in cold-rolled NiTi plates have a thickness of about 1 µm for manufacturing reasons. At typical heat-treatment temperatures between 400 and 600 °C, the oxygen further diffuses into the SMA component and changes the chemical composition by various reactions with Ti atoms and, thus, the shape memory properties. This problem is solved by removing the oxide layer with an etching solution, e.g., an aqueous solution of nitric acid and hydrofluoric acid.

Finally, the conditioned and sequentially manufactured SMA components are joined to an actuator using suitable assembly and bonding technologies. Only those bonding technologies may be applied that do not cause any large thermal loads of the SMA components. Due to the high stresses and strains occurring during operation, the mechanical connections must be sufficiently stable. Suitable bonding technologies are, e.g., welding processes, such as laser and ultrasonic welding; and, with certain limitations, adhesive bonding and fastening processes without any temperature effect, e.g., screwing, clamping, or pressing.

4.5 SMA Microactuators

First attempts to introduce SMA materials in miniature devices and systems have been based on miniaturization of conventional SMA semi-products. Typical examples are miniature grippers [234, 235], joint mechanisms [236–238], endoscopes [239, 240] and fibrescopes [241]. In this approach, the actuator components are fabricated individually by conventional machining methods and assembled later on. The results give interesting insights into the feasibility and functionality of miniature SMA devices. However, the approach is not workable in microsystems technology due to technological and economic reasons.

For the design of SMA microactuators, basic aspects of microsystems technology, as well as specific properties of SMAs have to be taken into account:

- Simple geometries: In order to facilitate fabrication and enable a large degree of miniaturization, microactuator designs require simple geometries.
- Multifunctionality: Preferably many functions should be performed by preferably few components.
- Monolithic designs: In the ideal case, all functional units of a microactuator are integrated in one piece of material. Thus, the microactuator can be realized in one process step.
- Design optimization: For optimum use of the shape memory effect, the mechanical and thermal optimization criteria (O1)–(O6) introduced in Sect. 4.3 have to be fulfilled.
- Batch fabrication: Following the procedure in microelectronics, SMA microactuators should be fabricated largely in parallel by applying standardized technologies for micromachining and integration. Movable structures may be realized, for instance, by a sacrificial technology, lift-off, or alternative processes like partial removal of the substrate. Due to the special properties of SMAs, the technologies demanded for realization of SMA microactuators are only partially available today.

Introduction of SMAs in microsystems technology started at the beginning of the 1990s [12, 14, 15]. From a technological point of view, two main directions of microactuator development may be distinguished:

- Hybrid integration of SMA components fabricated separately from the final microsystem. Generally, hybrid integration technologies may be applied to SMA plates, foils, or thin films. This approach is pursued in the framework of the development activities presented in Chaps. 5 and 6.
- Monolithic integration of SMA components based on deposition and micromachining processes using silicon technologies. This procedure will be addressed in the following section.

4.5.1 Monolithic Integration of SMA Microactuators

Early developments concentrated on the realization of simple geometries like microbeams [242, 243] and springs [12] in order to demonstrate the feasibility of SMA microcomponents and to explore the compatibility with the fabrication processes of other actuator components. Based on these developments, more complex microactuators have been realized using monolithic integration. In all cases, the base materials are sputtered SMA thin films.

4.5.1.1 Overview

Table 4.3 gives an overview of some SMA microactuators fabricated in silicon technology by monolithic integration.

Table 4.3. Examples of SMA microactuators fabricated by monolithic integration

Shape memory effect	Functional mechanism	Application	Ref.
One-way effect	Electrical actuation, work production with passive biasing mechanism	Microvalve	[71]
One-way effect	Electrical actuation, work production with antagonistic reset mechanism	Micropump	[244]
One-way effect	Electrical actuation, constrained recovery	Microclipping device	[245]
One-way effect	Thermal actuation, work production with passive biasing mechanism and constrained recovery	Microgripper	[246]

One of the first applications of SMA thin films was the microvalve, which was developed in silicon technology by Johnson et al [70–73]. Fig. 4.13 shows a schematic of the valve setup [71]. The valve consists of three Si layers. The top layer is micromachined to form a spring. The middle layer contains a poppet supported by ribbons of NiTi. The valve seat is located in the bottom layer.

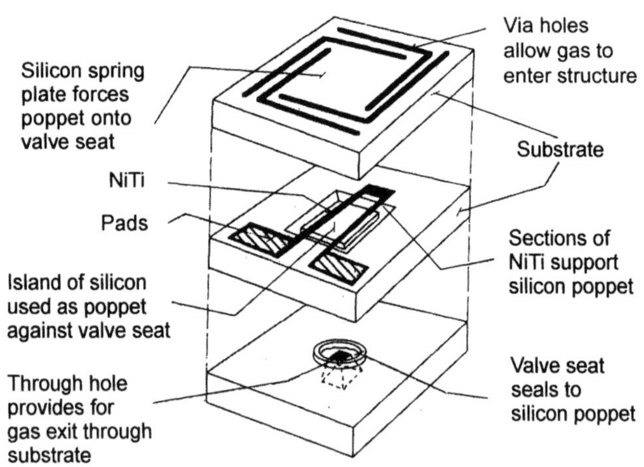

Fig. 4.13. Schematic of SMA microvalve fabricated in silicon technology by monolithic integration [71]

NiTi ribbons and the Si spring form an actuation system with a passive biasing mechanism. At room temperature, the microspring forces the poppet against the valve seat to close the valve while the ribbons are stretched. In this condition, the force of the microspring determines the maximum controllable pressure difference. Direct heating of the ribbons with electrical current causes the ribbons to contract, pulling the poppet back into the plane of the middle Si layer.

A reciprocating micropump has been developed by Benard et al [244]. A configuration comprising two antagonistic TiNi actuators is used to perform cyclic motion, as sketched in Fig. 4.14. The stroke of the actuator is increased as the heated SMA device pushes against the more compliant SMA device in martensitic condition. This results in less of the actuator's force being used to overcome the antagonist, delivering more force for pumping. Passive check valves are integrated to ensure unidirectional flow through the pump chamber.

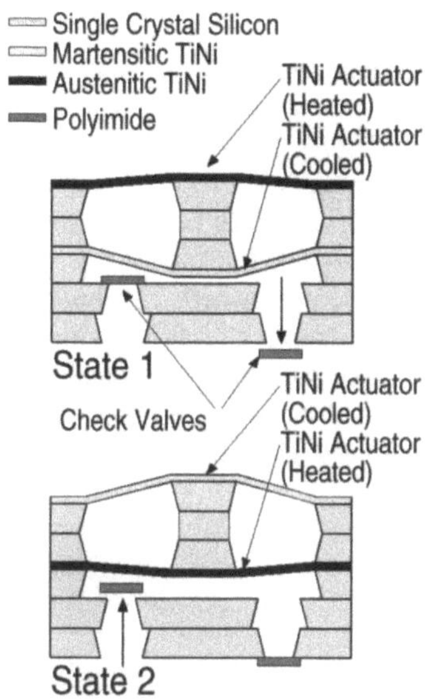

Fig. 4.14. Schematic of micropump fabricated in silicon technology by monolithic integration. In state 1, the pump chamber is compressed, forcing fluid out of the right check valve; in state 2, the pump chamber is expanded, drawing fluid in through the left check valve [244]

Takeuchi et al have developed a microclipping device for insect neural recording [245]. The structure and mechanism of the microclipping device are sketched in Fig. 4.15. The device consists of a clipping structure with folded beam shape and two electrical probes with single beam shape. By heat treating the probes in constrained condition, as illustrated in Fig. 4.15 (2), a three-dimensional shape is memorized. Before use, the beams are manually deflected using a micromanipulator to obtain a sufficiently large opening gap for insertion of the clipping object, see Fig. 4.15 (3).

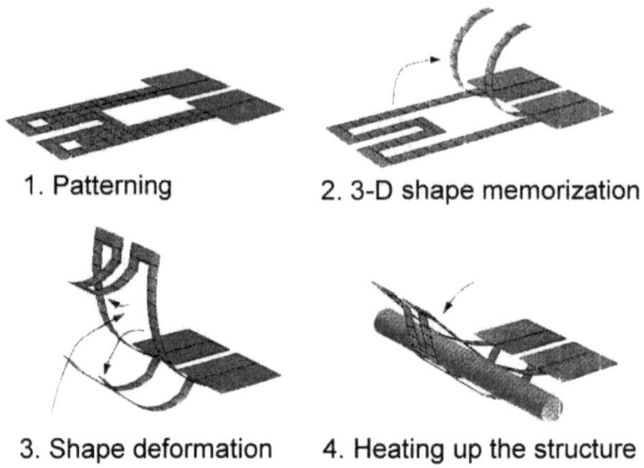

Fig. 4.15. Schematic of microclipping device fabricated in silicon technology by monolithic integration [245].

The device is used for clipping a nerve cord, e.g., of an insect. By electrical heating, the probes and clipping structure bend around the nerve cord in order to fix and simultaneously electrically contact it. In this condition, measurements of electrical potential difference are performed. The high elasticity of the SMA thin film structure also allows nondestructive tests on living insects [245].

A microgripper has been developed by Lee et al [246]. As illustrated by the three-dimensional schematic in Fig. 4.16, the microgripper consists of two micromachined Si wafers with sputtered SMA thin films on the top and bottom.

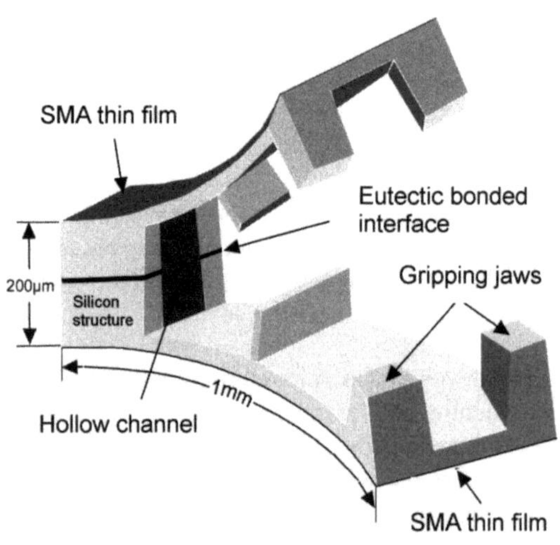

Fig. 4.16. Schematic of microgripper fabricated in silicon technology by monolithic integration [246]

The micromachined Si layers are used as passive biasing springs. A pushing pad is integrated to assist in releasing the gripped object by pushing forward as the gripper is opened. By deposition of the thin film near 500°C, a tensile thermal stress is created as it is cooled after deposition. Below the martensitic transformation temperature, the thermal stress can relax. Thus, the SMA thin film is stretched and the gripping jaws are closed. Actuation occurs by the recovery of tensile residual stress when heating the SMA film. In this case, the gripping jaws are opened by the contracting SMA film.

4.5.1.2 Design

The presented SMA microactuators fabricated by monolithic integration consist of SMA beam or bridge structures with rectangular geometry. The simple geometry allows analytic estimates of the mechanical properties, as for conventional SMA actuators. However, designs of rectangular geometry exhibit heavily inhomogeneous stress distributions, which may cause crack initiation and fatigue.

The design of thermal properties of the SMA microactuators is less critical as in the case of SMA macroactuators since the large surface-to-volume ratios favor a good convective heat exchange with the environment and, thus, a homogenization of temperature profiles. Due to the low thermal mass of SMA microactuators, only small amounts of heat need to be supplied and removed during operation. Consequently, power requirement and heat transfer times are strongly reduced. By using substrates with good thermal conductivity, like, e.g., Si substrates, excess heat is removed sufficiently fast. Depending on the thermal coupling between SMA microactuator and substrate, however, selective heating of the SMA microactuator may become difficult.

Concerning the design criteria (O1)–(O6) presented in Sect. 4.3 the following conclusions may be drawn:

(O1)–(O3): The option of almost arbitrary lateral shaping of SMA microcomponents has not been used for design optimization so far. Thus, inhomogeneous stress profiles are present, which may reduce the work output and possibly the lifetime.
Consequent implementation of the optimization concepts presented in Sect. 4.3 makes use of the design freedom with respect to the lateral shape of SMA microcomponents. In this case, however, the design necessitates the employment of FEM methods.

(O4)–(O6): The favorable thermal properties of SMA microcomponents facilitate thermal optimization. In most cases, adjustment of thermal coupling to the substrate is sufficient.

4.5.1.3 Fabrication Technologies

Starting material for fabrication of the presented SMA microactuators are sputtered and subsequently heat-treated NiTi-based thin films. Various technologies have been developed for micromachining, packaging, and interconnection. Table

4.4 summarizes the main process steps for fabrication of the SMA microactuators listed in Table 4.3.

Fig. 4.16 illustrates, for instance, the process sequence for the fabrication of the middle layer of the SMA microvalve [73]. First, a wafer is wet-etched from one side to create wells. This etching leaves thin Si membranes. Each membrane supports a Si island, which becomes the poppet of the microvalve. Then, the NiTi thin film is sputter-deposited, heat-treated, and patterned by wet-chemical etching. Subsequently, the remaining Si is removed under the NiTi ribbons. The wafer is finally separated into dies, each containing a single microactuator. These dies are assembled with corresponding dies of the top and bottom layer.

The main specifications of the SMA microvalve are summarized in Table 4.5.

Table 4.4. Process steps for the fabrication of various SMA microactuators by monolithic integration

SMA microactuator	Process steps	Ref.
Microvalve	Si bulk micromachining Magnetron sputtering of NiTi Heat treatment Chemical micromachining of NiTi Dicing Electrical contacting	[71]
Micropump	Si bulk micromachining Magnetron sputtering of NiTi Heat treatment Dicing Electrical contacting	[244]
Microgripper	Si bulk micromachining Precision sawing Eutectic wafer bonding Magnetron sputtering of NiTiCu Heat treatment Dicing	[246]
Microclipping device	Sacrificial layer technology Magnetron sputtering of NiTi Chemical micromachining of NiTi Deposition and micromachining of polyimid Dicing Heat treatment 3-D shape memorization Selective removal of substrate Electrical contacting	[245]

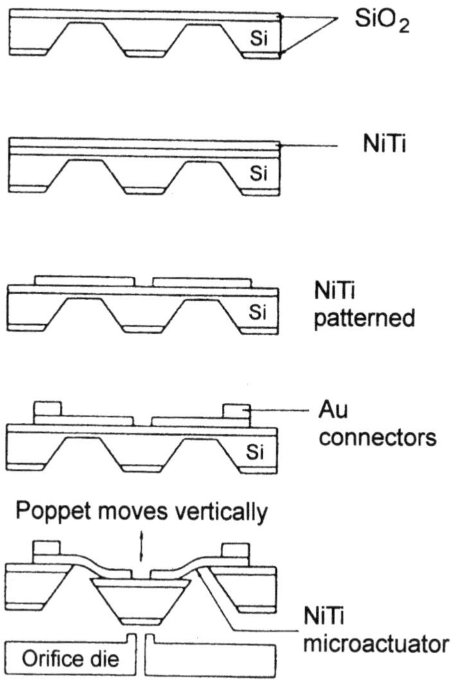

Fig. 4.17. Process sequence for the monolithic fabrication of the middle layer of the SMA microvalve [73]

Table 4.5. Specifications of the SMA microvalve fabricated by monolithic integration in Si technology. The abbreviation n.c. denotes normally closed

Type	Size (mm^2)	Max. flow / max. pressure difference (N_2 Gas)	Switching time (ms)	Power / voltage (DC)	Ref.
Si, n.c.	10 x 5	800 sccm / 670 kPa	50	< 0.3 W / < 3 V	[73]

A similar process sequence has been used for monolithic integration of the micropump shown in Fig. 4.14. The pump housing is fabricated by bulk Si micromachining. By this process, thin diaphragms of Si are left for subsequent sputter-deposition of TiNi thin films. After deposition and annealing, the Si diaphragms are etched to release the TiNi films. Then, after dicing, two TiNi devices are bonded together with a spacer between the bosses. Finally, the check valve wafers and the assembled pump chamber are bonded to complete the pump fabrication.

The processes for monolithic integration of the SMA microgripper shown in Fig. 4.16 comprise a combination of precision sawing and bulk silicon micromachining. The hollow channel between the gripping jaws is fabricated by lithography and wet-etching. For microfabrication of channels perpendicular to the hollow channel, precision sawing is chosen to avoid anisotropic etching limitations of (110) silicon. After sawing, two identical wafers are bonded by a selective eutectic bonding process, which generates a bond at the base end, but allows gaps to be opened at the gripping end. Thus, freely movable gripping jaws are obtained. Sputter deposition and heat treament are performed after processing the Si wafers. Thin films of NiTiCu are sputter-deposited on the top and bottom side of the bonded pair of wafers. As the thin films are processed sequentially, one SMA film is annealed twice.

Fabrication of the microclipping device shown in Fig. 4.15 deviates from the scheme of Si bulk micromachining. Instead, surface micromachining based on a sacrificial layer of Ti has been developed. TiNi thin films are sputter-deposited and patterned by wet-etching. Free-moving beams are created by removing the sacrificial layer. After dicing, heat treatment is performed by using an infrared lamp annealer. A protection layer of polyimide is introduced on top of the thin film to avoid peeling off the substrate during heat treament. Since the application of clipping nerve cords requires a three-dimensional memory shape, a special shape memorization technique has been developed. A bond wire forming a small loop across the SMA microstructure on the substrate is used to constrain the clipping structure during annealing. In order to obtain a large work space of the microelectrode, the substrate under the electrode is partly cut off.

The described processes of monolithic integration allow parallel fabrication of micromechanical structures of Si, as well as SMA thin films following the procedures of microelectronics. However, this approach is associated with a number of constraints concerning the choice of process parameters:

- High annealing temperatures cause problems of adhesion of the SMA thin films, as well as other microstructures on the substrate.
- Temperature-sensitive components on the substrate force heat treatment of the SMA thin films to be performed at reduced temperatures.
- The memory shape is mainly restricted to planar geometries or geometries bent out of the plane of substrate.
- Wet-chemical micromachining of the SMA thin films requires the use of aggressive etching solutions that require measures for protection of other components on the substrate.
- Monolithic integration is restricted to SMA thin films of thicknesses below about 20 μm fabricated by deposition technologies.

These restrictions can be avoided by the separate processing of the SMA microcomponents and subsequent hybrid integration.

4.5.2 Hybrid Integration of SMA Microactuators

Through hybrid integration, various SMA materials such as sputter-deposited thin films, cold-rolled foils, or melt-spun ribbons may be combined with a substrate. Meanwhile, NiTi foils have been fabricated by cold-rolling in all thicknesses down to 5 μm [19,20]. Thin SMA ribbons of various chemical compositions have been produced by melt spinning [247]. Due to restrictions in production, the width of the ribbons is limited to a few millimeters. Typical thicknesses are between 20 and 50 μm. As the SMA materials are produced independently of a substrate in the cases of cold rolling and melt spinning, hybrid integration is essential for the realization of SMA microactuators.

Depending on the application, the use of SMA ribbons, foils, and plates may offer some advantages, which are summarized in the following:

- Since the materials are produced by melting and casting, approved material properties are available for use in microactuators and microsystems.
- The thickness of these materials, in particular of NiTi foils and plates, can be adjusted in a large range.
- For thicknesses above 30 μm, microstructures of sufficient aspect ratio can be realized that allow bending motions within the plane of substrate.
- By using bending motions in microstructures of sufficient thickness, both large displacements and large forces can be generated.

These advantages have been recognized at the Forschungszentrum Karlsruhe in an early stage and implemented in the development of SMA microactuators [200, 243]. Some of the development activities will be presented in detail in Chaps. 5 and 6.

4.6 Development Trends

Conventional macroscopic SMA actuators have gained only minor importance compared to actuators based on other actuation principles, even though they provide large work outputs. The main drawbacks are high material costs and a number of functional disadvantages like long time constants and low thermodynamic efficiency. Due to favorable scaling behavior, however, the disadvantages become less severe upon miniaturization. At the same time, the advantages of SMA actuators, like large work output, gain in importance due to the partly unfavorable scaling behavior of other actuation principles. Therefore, SMA microactuators have a considerably larger application potential than conventional SMA actuators in macroscopic dimensions.

Previous developments of SMA microactuators highlight the substantial progress in the development of material properties and technical feasibility. The examples presented already exhibit promising specifications. For a breakthrough in SMA microactuator applications, however, further developments are required and further challenges have to be met. As listed in Table 4.6, current topics of research and development may be classified into various fields like materials research, modeling and simulation, design optimization, micromachining, packaging technologies, and control strategies.

Table 4.6. Topics of research and development in the field of SMA microactuators

Fields of R&D	Topics of R&D
Materials research	Stability of shape memory effect Range of phase transformation temperatures Hysteresis width Fatigue effects Brittleness Material anisotropy Biocompatibility
Deposition technologies	Magnetron sputtering Electron-cyclotron-resonance sputtering Ion beam sputtering Laser ablation Flash evaporation Melt spinning
Modeling and simulation	Multivariant micromechanical modeling Macroscopic phenomenological modeling Finite element simulation Coupled multilevel simulation
Design optimization	Computer aided optimization (CAO) Stochastic derivative-free methods: - evolutionary strategies - genetic algorithms
Micromachining	Chemical Etching Electrolytic photoetching Lift-off technology Laser cutting Laser ablation Spark erosion
Packaging and interconnection technologies	Adhesive bonding Laser welding Laser soldering Gap welding
Driving and control strategies	Adaptive systems - PID - fuzzy logic - neural networks Inverse control

Current challenges in the field of SMA microactuators are:

- The application potential of SMA microactuators is determined by available SMA materials. Therefore, the breadth of material properties needs to be further increased, allowing precise adjustment, e.g., of phase transformation temperatures or of hysteresis width in a large range.
- SMA microactuators can be produced with a large variety of lateral shapes. Therefore, tools for field simulation have to be developed taking into account all aspects of shape memory behavior, as well as the coupling of different

physical properties. For this purpose, an appropriate trade-off between simulation time and accuracy has to be found.

- The relatively complex material properties of SMAs are one reason why the design of SMA actuators still belongs to the domain of specialists. Therefore, computer-based design tools, including design optimization and material databases, are required for broad use.
- For further miniaturization and integration of additional functions in SMA microactuators, more technologies have to be to developed that are compatible with available processes. Increasing complexity demands increasingly three-dimensional structures. Thereby, the need for hybrid integration technologies increases in particular.
- The small dimensions of SMA microactuators require novel test routines for failure-free measurement of functional parameters with sufficient precision. The realization of integrated test routines itself requires microtechnologies, which must be taken into account in the design.
- Suitable strategies for driving and control need to be developed, in particular, for positioning tasks. Simple procedures are based on the use of additional microsensors in a closed-loop control scheme. Compact solutions may be realized through the simultaneous use of actuation and sensing capabilities of the transducer (smart actuation) [9]. By digital signal processing, complex control strategies based, e.g., on fuzzy logic or neural networks are also realizable cost-effectively. Alternative procedures are based, e.g., on inverse models, which rely on theoretical predictions of actuation behavior [248].
- SMA microactuators need to be implemented in actuation systems and system families consisting of various microactuators, microsensors, and control units that are adjusted with respect to each other. The resulting increase of functionality will have a strong impact on the breadth of applications.
- By developing modular microactuator designs with standardized interfaces, microsystems with flexible composition that can be adapted to application-specific requirements can be realized cost-effectively.

5 SMA Microvalves

Cui bono? (What is the good of it?)

–Marcus Tullius Cicero

In terms of identifying applications of SMA microactuators the field of microfluidics appears to be particularly promising. An approach to the development of SMA microvalves is described in the following. It is based on hybrid integration of micromachined SMA foils and thin films with optimized lateral shape. Recent publications on this subject can be found in Refs. [186, 199, 206, 220, 249–260].

At first, the basic setup and functionality of the valves will be explained, the valve design will be introduced, and an overview of different development directions will be given. The specifications of different valve variants will be compared with the specifications of other microvalves summarized in Chap. 2 and discussed with respect to possible applications.

5.1 Valve Concepts

5.1.1 Setup and Operation Principle

In the simplest case, the SMA microvalves considered here are 2/2 normally open valves with two fluidic connections for the switching of fluid flows in two opposite directions. The basic setup is sketched in Fig. 5.1. The main valve components are an upper and a lower housing with integrated valve seat and fluidic connections, a membrane, a spacer, and an SMA microactuator.

The underlying ideas of the chosen setup may be summarized as follows:

- The valve housing is divided in two halves. The lower half forms the fluid chamber, the upper half the actuation chamber. The partitioning of the two halves allows simple fabrication of housings based on polymers using mass-fabrication methods such as hot embossing or injection molding. In principle, different materials may be selected for both halves in order to satisfy special demands on the actuation and fluid chamber, respectively.
- A spatial separation between fluid and actuation chamber can be easily realized by integrating a membrane between the upper and lower housing, in particular, when the fluid and actuation chambers should be kept isolated from each other. In this case, the membrane is used at the same time for sealing the valve seat.

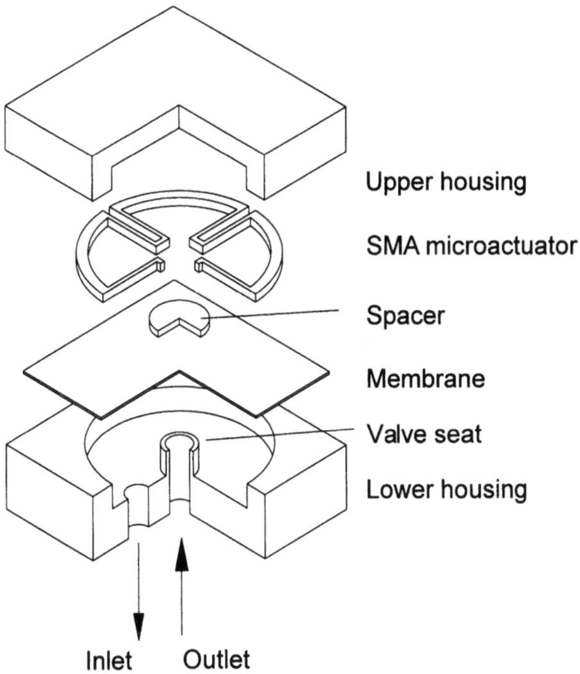

Fig. 5.1. Exploded view of a 2/2 normally open SMA microvalve

By micromachining the membrane, a pressure difference between the fluid chamber and actuation chamber can be compensated. If only a sealing function is required, the membrane may be replaced by a sealing plate or plunger. The fabrication and integration of a membrane in between two polymer housings can be performed in batch mode [261].

- In the actuation chamber resides a micromachined SMA actuator that generates forces and displacements in the direction perpendicular to the valve seat by using the one-way shape memory effect. The SMA microactuator is designed for direct heating by an electrical current. The actuator geometry consists of a circular arrangement of beam elements or of bridges with planar memory shape. Due to its monolithic design, fabrication of SMA microactuators can be performed in batch mode by micromachining planar SMA materials such as plates, foils, or thin films.
- The maximum strain and stress in the SMA microactuator are adjusted by using a spacer. Since the thickness of the spacer is arbitrary, the maximum stroke and flow, as well as the maximum controllable pressure can be varied in a wide range. The integration of the spacer may be performed by pick and place technology.

The operation principle of the valve shown in Fig. 5.1 is illustrated in Fig. 5.2. In martensitic condition, already a small pressure difference between the inlet and outlet is sufficient to deform the membrane and SMA microstructures in the direction perpendicular to the valve seat. Thus, the valve is in a normally open condition, see Fig. 5.2(a). By heating the SMA microactuator above the austenitic phase transformation temperature, the planar memory shape is recovered. Consequently, the membrane is pressed onto the valve seat, whereby the valve is closed. The reset function in the unheated condition is accomplished by the pressure difference between the inlet and outlet. By simple modification of the setup, normally closed valves can also be realized.

The valves are produced by hybrid integration of the valve components, which are fabricated largely by parallel processes. For integration of the SMA microactuators, a specially developed hybrid integration technology is used that has proven itself particularly in combination with very small and delicate structures. The mechanical interconnection of the valve components is made using different adhesive bonding technologies. For testing purposes, electrical connections are realized by wire bonding. For industrial applications, other technologies like laser welding or integration of conduction strips are of interest. Conduction strips are anisotropically conductive foils, which are utilized industrially, e.g., for the contacting of displays. They consist of a rubber-like elastic strip with embedded conducting pins of Au that provide an anisotropic conducting connection.

5.1.2 Design

The SMA microvalves that will be presented in the following result from research and development activities. Therefore, their design will not begin with specifications aiming at certain performance data. Instead, the starting point is the available SMA materials and the basic construction elements for the development of SMA microvalves whose functional properties are of interest in a later development stage.

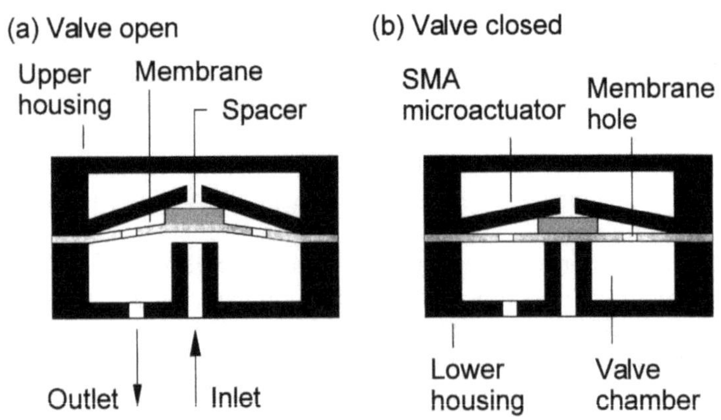

Fig. 5.2. Operation principle of a 2/2 normally open SMA microvalve

The valve design comprises the components for actuation (SMA microactuator, spacer, biasing spring where necessary), as well as the fluidic components (valve seat, membrane, fluidic inlet and outlet). Generally, the mechanical and thermal properties of the components for actuation are closely coupled with the fluidic properties of the valve. A thermally induced motion of the SMA microactuator causes a change in flow. This results in a change of pressure and convective cooling and, thus, retroacts in a complex way on the SMA microactuator. An experimental determination of spatial and time-dependent flow profiles is not possible due to the small dimensions. Thus, a detailed investigation of the mechanical, thermal, and fluidic properties demands coupled multilevel simulation.

However, the valves considered here feature several properties, which allow for a simplified treatment:

- In austenitic condition the SMA microactuator acts like an elastic spring. Consequently, a simplified description of the balance of forces between SMA actuator and fluid is adequate.
- In open and closed conditions only the states of complete phase transformation in the SMA actuator are of interest. In this case, pure phase states (austenite, martensite, or R-phase) and, thus, reproducible strain and stress states are present, which allow a linear description based on an effective E-modulus even for nonlinear stress-strain characteristics.
- The flow-pressure characteristic is mainly determined by the annular gap between valve seat and membrane, allowing a simplified analytic treatment.
- The dynamics of the SMA actuator are determined by the heat-transfer times upon heating and cooling. In stationary equilibrium between heating and cooling processes the valve performance can be considered time-independent. In particular, inertia effects are negligible.

5.1.2.1 Mechanical Properties

In first approximation, the mechanical behavior of the valves in open and closed conditions can be approximated analytically. Depending on the used basic construction element, the elastic theory of cantilevers or bridges is thereby considered. Spatially resolved stress and strain profiles in the SMA microactuators are determined by FEM simulations. In order to make optimum use of the shape memory effect, the lateral shape of the SMA microactuators is optimized. The maximum work density of the microvalves is determined by the mechanical load limits of the SMA microactuators and of biasing elements where applicable.

5.1.2.2 Thermal Properties

The thermal behavior of the SMA microactuator is determined by the electrical heating power and the cooling mechanisms of heat conduction and convection in the valve. In the temperature range below 200 °C considered here, heat radiation plays an insignificant role. The spatial distribution of heat sources and sinks in the valves requires a thermal FEM simulation. In first approximation, the coupling between the mechanically determined gap size at the valve seat and convective cooling of the SMA microactuator is described by an effective heat-transfer coefficient. For minimization of the opening times of the valves, the heat-transfer rate of the SMA microactuator upon cooling must be optimized. For this purpose, a sufficiently dimensioned heat sink may be provided, which is connected with the SMA microactuator via a heat conductor.

5.1.2.3 Fluidic Properties

For the analytic description of the flow-pressure characteristic in the annular gap between valve seat and membrane, a number of specifics have to be taken into account, which result from the small dimension of the gap length L in flow direction compared to the hydraulic diameter D_h. Macroscopic flow conditions would be present in the case $L/D_h>70$, whereby in this case a transition from laminar to turbulent behavior occurs at a critical Reynolds' number of about 2300 [262]. The microvalves considered here are characterized by an L/D_h-ratio well below 70. This results in the following consequences [263]:

- At small L/D_h-ratio no turbulent flow can form.
- A critical Reynolds' number $Re_t = C/\zeta \cdot L/D_h$ exists, above which a transition between laminar flow and not fully developed flow occurs, where the pressure drop is dominated by inertial losses. The parameter C is a friction coefficient, which is determined to 96 for rectangular cross sections. For $L/D_h<1$, the parameter ζ is about 2.6.

For the microvalves considered here, the L/D_h-ratio is in the range 0.2–2 and the critical Reynolds' number Re_t is between 8 and 75. However, the corresponding pressure and flow ranges result in considerably higher Reynolds' numbers compared to Re_t. Consequently, not fully developed flows with dominating inertial losses have to be assumed. In this case, the flow-pressure characteristic can be simulated by the following approximation [263]:

$$Q = \frac{A}{2\sqrt{\zeta}} \sqrt{\frac{2}{\rho} \frac{p_{in}^2}{p_{out}}} \qquad (5.1)$$

Eq. 5.1 can be used for diatomic gases at isothermal conditions, where the ratio between the pressures at inlet and outlet p_{in}/p_{out} satisfies $p_{in}/p_{out} > 2$. For diatomic gases and $p_{in}/p_{out} < 2$, as well as incompressible liquids the following approximation holds:

$$Q = \frac{A}{\sqrt{\zeta}} \sqrt{\frac{2}{\rho} \Delta p} \qquad (5.2)$$

Thereby, A and ρ denote the cross-sectional area of the flow path and the fluid density at output. The pressure drop Δp denotes the difference between the pressures at inlet and outlet p_{in}-p_{out}. The coupling between flow Q and the stroke of the valve membrane above the valve seat is described here by the cross-sectional area A, which is calculated from the stroke of the actuator Δz and the mean circumference of the valve seat.

In microfluidic flows, further specifics may possibly occur that lead to deviations from simple flow models:

- Particularly in gases the mean free path may reach the dimensions of the flow channel.
- Effects of surface tension and gas bubbles play a role in liquids.

These effects will not be considered within the scope of the development activities described here.

5.1.3 Overview

Table 5.1 gives an overview of various SMA microvalves developed at Forschungszentrum Karlsruhe. In general, the operating modes "normally open" and "normally closed" are distinguished, leading to different valve designs. Further distinctions result from the use of various SMA materials that require different technologies for valve fabrication. So far, rolled SMA thin plates with thicknesses between 70 and 100 µm, rolled SMA foils of 20 µm thickness, as well as sputtered SMA thin films with characteristic thickness of 10 µm have been utilized. A further distinctive feature follows from the use of the valves either for gas or for liquid operation. Partly, the valve variants are organized in subvariants, which have been fabricated for investigation of further design parameters.

Table 5.1. SMA valve variants. The abbreviations n.o. and n.c. indicate the operation modes "normally open" and "normally closed"; *d* denotes the thickness of the materials

	Variant 1	Variant 2	Variant 3	Variant 4	Variant 5
Material of SMA microactuator (one-way effect)	SMA thin plate d = 100 µm	SMA foil d = 20 µm	SMA foil d = 20 µm	SMA thin film d = 10 µm	SMA thin plate d = 90 µm
Material of biasing element (pseudoelastic effect)	-	-	-	-	SMA thin plate d = 70 µm
Operation mode	n.o.	n.o.	n.o.	n.o.	n.c.
Medium	N_2 gas	N_2 gas	water	N_2 gas	N_2 gas

5.2 Material Properties

The materials used for valve fabrication differ in chemical composition, as well as in the manufacturing method. In the following, the main thermal and mechanical material properties are summarized. Typical material characteristics can be found in Sects. 3.4.2–3.4.5.

5.2.1 Rolled SMA Thin Plates and Foils

Except for valve variant 4, rolled thin plates and foils of NiTi are used for valve fabrication whose thickness enters into the valve design as an arbitrary parameter. The materials have been treated to show the one-way effect. For the normally closed valves, furthermore, NiTi thin plates with pseudoelastic effect are used as a biasing element. Details on the fabrication of thin plates and foils can be found in Sect. 4.4.3.

5.2.1.1 NiTi Specimens with One-Way Effect

NiTi thin plates of 100 μm thickness and foils of 20 μm thickness have been received from different manufacturers. The alloy composition has been determined by chemical analysis. Both the Ni and Ti content of the thin plates is 50 at.%. The foils have a somewhat smaller Ti content of 49 at.% and a correspondingly higher Ni content of 51 at.%. The oxygen content is below 0.1 at.% in both cases.

At first, the fabrication-induced oxygen layer is removed. Subsequently, a heat treatment of the SMA thin plate is carried out in a pipe furnace in planar configuration. The heat treatment is performed at a temperature of 823 K for 10 minutes. After the heat treatment, the furnace is flushed with nitrogen in order to accelerate the cooling process and to reduce the formation of precipitates. The memory shape is determined by the heat treatment. Furthermore, the phase transformation temperatures are adjusted. These temperatures are determined by DSC analysis, as well as by electrical and mechanical measurements using test samples of double-beams.

The transformation temperatures of the specimens determined by DSC analysis are summarized in Table 5.2. Upon cooling, a two-step phase transformation of the sequence austenite → R-phase → martensite is observed in both cases. Upon heating, only a single-step phase transformation of the martensitic phase occurs directly to the austenitic phase. The martensitic phase transformation primarily takes place below room temperature. Consequently, only a single-stage R-phase transformation occurs by thermally cycling the materials at room temperature using an electrical current. This behavior is also observed in the case of mechanical loading as long as the maximum strain remains below about 0.6% [140]. At higher strains, SIM formation also occurs. At about 0.8%, the yield limit for plastic deformation is reached. Due to R-phase condition, this limit corresponds to a stress between 60 and 90 MPa, depending on the loading direction. In austenitic condition, the yield limit for plastic deformation is exceeded for a stress between 250 and 400 MPa according to the thermo-mechanical treatment of the material.

Mechanical investigations on test samples of double beams at various temperatures allow a determination of the E-modulus in austenitic condition E^A and of a strain-dependent effective E-modulus in R-phase condition $\overline{E}^R$. An analysis of the measured displacements using an elastic beam model results in the E-moduli summarized in Table 5.3.

Table 5.2. Chemical composition and transformation temperatures of NiTi thin plates and foils with one-way effect

Material	M_s (°C)	M_f (°C)	R_s (°C)	R_f (°C)	A_s (°C)	A_f (°C)
NiTi-50 at.%, d = 100 μm (cold work 20%, heat treatment 823 K for 10 min)	23	14	43	38	55	64
NiTi-49 at.%, d = 20 μm (degree of cold work unknown, heat treatment 723 K for 30 min)	16	-20	44	25	38	60

Table 5.3. E-moduli of NiTi thin plates and foils with one-way effect. The determination of the E-moduli has been performed at the temperatures $T \approx A_f$ (E^A) and T = 23 °C ($\overline{E}^R$). ε_{max} denotes the maximum strain in the sample

Material	E^A (GPa)	$\overline{E}^R$ (GPa)
NiTi-50 at.%, d = 100 μm	53	12 (ε_{max} = 0.8 %)
NiTi-49 at.%, d = 20 μm	55	23 (ε_{max} = 0.4 %)

By cold rolling, a preferred orientation is imprinted into the material, which leads to an orientation dependence of tranformation strain. As long as the shape memory effect is only used on the basis of a pure R-phase transformation, the influence of rolling direction on the mechanical properties of the material can be neglected.

5.2.1.2 NiTi Specimens with Pseudoelastic Effect

NiTi thin plates displaying a pseudoelastic effect are received from the supplier with a thickness of 90 μm with shiny etched surface and planar memory shape. The mass contents are 56.6% Ni and 43.3% Ti. For this alloy, the end temperature of the austenitic transformation A_f determined by DSC analysis is about 10 °C. Thus, the pseudoelastic effect is caused by SIM formation.

5.2.2 Sputtered SMA Thin Films

In manufacturing SMA microactuators, thin films of the material system Ni-Ti-Pd-Cu are also used, which have been fabricated by magnetron sputtering. The thin films are sputtered onto a polished alumina substrate, which has been coated by a sacrificial layer of Au. Alumina is chemically inert with respect to the used etching solutions and thermally stable above 800 °C. Between the alumina, sacrificial layer, and SMA thin film a good adhesion is achieved whereby the small difference of the coefficients of thermal expansion leads to a tolerable thermally induced stress during heat treatment. The stoichiometry of the targets produced by hot pressing is $Ti_{54}Ni_{46}$, $Ti_{55}(Ni_{65}Cu_{35})_{45}$, and $Ti_{54}(Ni_{40}Pd_{60})_{46}$. For a distance between the target and substrate of 50 mm, an Ar pressure of 0.4 Pa, and a sputtering power of 300 W, the deposition rate is between 7.5 and 10 μm/h, depending on the target composition. The as-deposited amorphous thin films are subsequently crystallized. For this purpose, an annealing step is performed in a tube furnace in high vacuum at 843 K for 45 min. Thereby, the thin films are clamped between Ti plates, which act as a getter material for residual oxygen and, thus, avoid the formation of an oxide layer.

The chemical composition of the thin films is determined by wavelength dispersive X-ray microanalysis (WDX). For all films, a reduction of the Ti content by 1.8–3.5 at.% is found compared to the target composition. The oxygen content is below 1 at.%. Details of the fabrication and characterization of the SMA thin films can be found in [208].

Transformation temperatures are determined by DSC analysis, as well as electrical and mechanical measurements on test samples with double-beam geometry. Table 5.4 contains integral chemical compositions of the thin films and transformation temperatures determined by DSC analysis.

The low martensitic transformation temperatures of the binary thin films indicate a Ni content above 50% in the matrix. Thus, the integral chemical composition does not reflect the matrix composition. Since the martensitic transformation occurs below room temperature, thermally cycling the thin films at room temperature by an electrical current induces a pure R-phase transformation. In contrast, ternary NiTiCu and NiTiPd alloys exhibit a martensitic transformation above room temperature.

The marterial system Ni-Ti-Pd is particularly suitable for adjusting the operation temperature of the microvalves. An increase of the Pd content allows a strong increase of the transformation temperatures of the NiTiPd thin films. However, the transformation temperatures are affected in addition by the annealing temperature [207]. In the present case, a relatively low annealing temperature is selected in order to avoid diffusion of the Au sacrificial layer used into the NiTiPd thin film. Consequently, relatively low transformation temperatures arise. The use of the present NiTiPd thin films enables valve operation at ambient temperatures up to 60 °C. For the adjustment of higher transformation temperatures, an alternative sacrificial layer technology is required.

Table 5.4. Integral chemical compositions and transformation temperatures of sputtered thin films

Material	M_s(°C)	M_f(°C)	R_s(°C)	R_f(°C)	A_s(°C)	A_f(°C)
(a) $Ti_{52}Ni_{48}$	-40	< -70	50	35	37	52
(b) $Ti_{53}Ni_{30}Cu_{15}$	36	19	-	-	35	52
(c) $Ti_{52}Ni_{20}Pd_{28}$	89	60	-	-	84	132

The E-modulus of the thin films in austenitic condition E^A is determined from load-dependent displacements of double-beam test samples. An analysis of the results using an elastic beam model yields: E^A = 65 GPa (NiTi), E^A = 100 GPa (NiTiCu), and E^A = 105 GPa (NiTiPd). The relative error of these values is about 20%, since the small film thickness gives rise to a strong influence of deviations from a constant thickness profile and of inaccuracies of thickness measurement.

5.3 Normally Open SMA Microvalves

In the absence of external heating power, normally open valves remain in an open condition. Due to the high compliance of the SMA microactuator in martensitic or R-phase condition, only a small pressure difference between inlet and outlet is sufficient to open the valve. Besides opening the valve, the pressure difference is used for resetting the SMA microactuator. Thus, no biasing element is needed.

5.3.1 Valve Designs

The design of SMA microvalves starts with several SMA materials and the basic construction elements fabricated thereof, such as "beams" or "bridges." The procedure for designing the SMA microactuators and valve components is based on the optimization criteria (O1)–(O6) discussed in Sect. 4.3.

5.3.1.1 Variant 1 with SMA Thin Plate Actuator

The underlying idea of valve variant 1 is the use of forces and displacements of SMA beam elements. Investigations on test structures in the thickness range of d=100 μm demonstrate that SMA beams are a suitable construction element that allows favorable combinations of large forces and displacements [185, 200]. For increase of force, several beams are arranged in a circle around the valve seat. The beams are interconnected serially in order to electrically heat all beams simultaneously. Electrical interconnection to the power supply is provided by contact pads. The design of the SMA microactuator is shown in Fig. 5.3. Fabrication has been made by laser cutting.

Fig. 5.4 schematically shows the load case for valve variant 1 in closed condition. The SMA beam is deformed by the spacer in the direction perpendicular to the valve seat. This results in a force at the contact points between the SMA beams and spacer directed perpendicular to the valve seat. In closed condition, the following force balance appears:

$$F^A = F_{\Delta p} + F_c \,, \tag{5.3}$$

where F^A, $F_{\Delta p}$, and F_c denote the force of the beams in austenitic condition, the pressure-induced force, and the closing force, respectively.

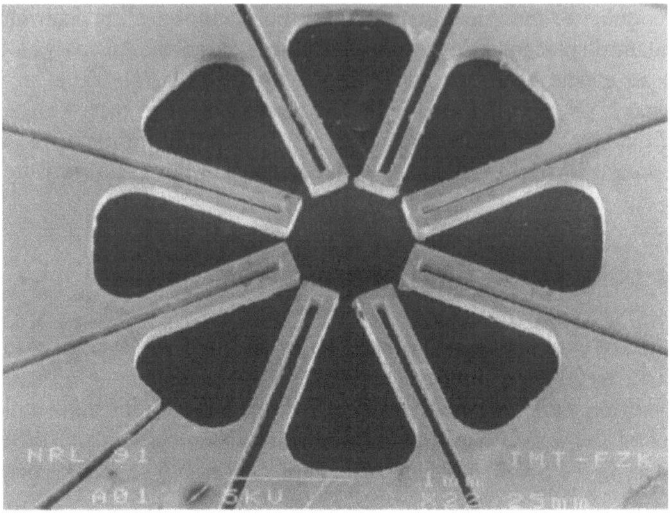

Fig. 5.3. SMA microactuator consisting of a circular arrangement of double-beam elements of constant width

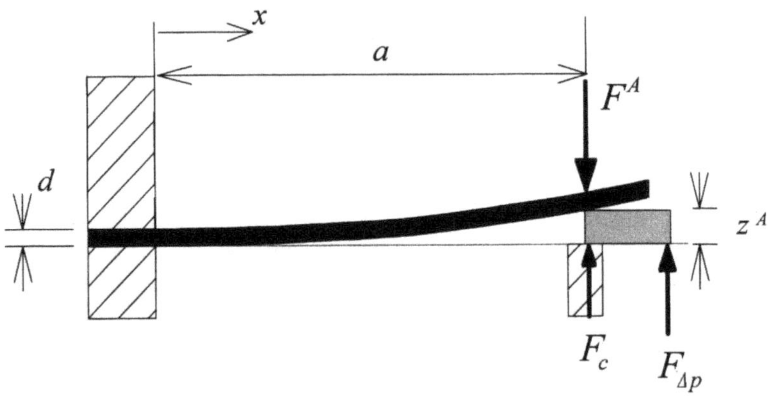

Fig. 5.4. Load case of a beam in valve variant 1 in closed condition

For a given force of the beams F^A the closing force decreases with increasing pressure-induced force until the force balance $F^A = F_{\Delta p}$ is finally reached and the closing force disappears. In this case, the maximum controllable pressure difference is present. At higher pressure-induced forces, the valve opens and remains in open condition independent of the phase state of the SMA beams. For a given pressure-induced force below the maximum controllable pressure difference, a phase transformation from the austenitic phase to R-phase causes a change of the force balance given by Eq. 5.3 in favor of the pressure-induced force. This causes the valve to open once the pressure-induced force outbalances the closing force. In open condition, the following force balance appears:

$$F^R = F_{\Delta p}\,. \tag{5.4}$$

For beams of constant width and thickness, a point load causes a stress peak near the clamping position. In this case, the shape memory effect is used only inefficiently. In order to fulfill the mechanical optimization criteria (O1)–(O3), the lateral width of the SMA beams *w(x)* needs to be optimized. In the present load case, optimization can be performed analytically. Since the SMA beams generate their maximum force in closed condition, the austenitic state is thereby considered.

The bending moment $M_b(x)$ is determined by the sum of pressure-induced force $F_{\Delta p}$ and closing force F_c:

$$M_b(x) = (F_{\Delta p} + F_c)(a - x)\,. \tag{5.5}$$

From Eqs. 4.21, 5.3, and 5.5, the optimized width profile results:

$$w(x) = \frac{6F^A(a-x)}{\sigma_{limit} d^2}\,. \tag{5.6}$$

The width of the beams increases linearly from the beam end to the clamping position. In this case, a homogeneous stress of σ_{limit} occurs along the surface on the top and bottom of the beam. The stress σ_{limit} is determined by the tolerable load limit of the material and by requirements on the maximum work density. For the R-phase transformation used, a stress limit in the austenitic state of 130 MPa is assumed in order to keep below the 0.8% strain limit. The required force of the beams in austenitic state F^A is determined by the maximum controllable pressure difference and the geometry of the valve seat.

Fig. 5.5 shows an SMA microactuator with optimized width profile of the beams. Fabrication has been done by laser cutting a rolled SMA thin plate.

Fig. 5.5. SMA microactuator consisting of a circular arrangement of double beams with optimized width profile

Theory suggests a width profile with zero beam width at the free end. The real version keeps a minimum beam width of 100 µm due to technological reasons. These regions remain below the load limit, which, however, has only a small influence on the force balance. The beam width is designed no larger than 600 µm in order to keep the heating current low. The minimum separation of the beams is limited by the cutting clearance of the used laser of about 50 µm. For a 4 mm-diameter valve chamber, these boundary conditions lead to a microactuator design with eight identical stress-optimized double beams.

The displacement of the SMA beams in austenitic condition is determined from elastic theory of trapezoidal beams [264]:

$$z^A = 4k \frac{F^A}{E_A w(0)} \frac{a^3}{d^3}. \tag{5.7}$$

The parameter k depends on the ratio of width $w(a)/w(0)$ and varies between 1 and 1.5. From Eq. 5.7, the thickness of the valve spacer can be calculated by inserting the required force of the beams F^A.

The displacement of the SMA beams in R-phase condition is determined in a manner analogous to Eq. 5.7 by using the corresponding force F^R and effective E-modulus $\overline{E}^R$. The stroke of the SMA actuator follows from the difference of the displacements z^A and z^R. The stroke of the membrane and the geometry of the valve seat determine the cross-sectional area of the flow path at the valve seat and, thus, the flow through the annular gap using Eqs. 5.1 and 5.2. In the case of pressure compensation between valve and actuation chamber, the force balance given by Eq. 5.4 no longer holds.

Valve variant 1 is fabricated in two subvariants, which mainly differ with respect to the width of valve seat. By comparing both subvariants, the influence of the width of valve seat on the sealing behavior in closed condition can be studied. The major design parameters are summarized in Table 5.5.

5.3.1.2 Variants 2 and 3 with SMA Foil Actuator

SMA microactuators of small thicknesses are particularly suitable for the realization of compact sizes and favorable dynamic properties. In the course of further miniaturization of the SMA microactuator the primary objective is to maintain the driving force as much as possible. Thereby, a decrease of displacement of the actuator is accepted. Upon thickness reduction of SMA beams, the requirement of an unchanging force cannot be fulfilled. Therefore, the concept of beams is dismissed for the case of SMA actuators of small thicknesses. Investigations on test structures with thicknesses in the range 10–20 µm demonstrate that SMA bridges can fulfill the objectives of "high force" and "moderate displacement" instead. As for the case of beams, a circular arrangement of several bridges is chosen in order to increase the force. Fig. 5.6 shows, for instance, an actuator design consisting of three SMA bridges that are electrically connected in series.

Table 5.5. Design parameters of SMA valves of variant 1

	Variant 1a	Variant 1b
Diameter of valve chamber	4 mm	4 mm
Outer diameter of valve seat	1 mm	1 mm
Inner diameter of valve seat	0.5 mm	0.9 mm
Dead volume	3.5 μl	1.4 μl
Diameter of inlet and outlet	0.5 mm	0.4 mm
Diameter of spacer	1.6 mm	0.7 mm
Thickness of spacer	60 μm	50 μm
Minimum beam width of SMA actuator	100 μm	100 μm
Maximum beam width of SMA actuator	550 μm	600 μm
Beam length	1.5 mm	1.75 mm
Number of beams	16	16
Thickness of SMA actuator	100 μm	100 μm
Membrane thickness	3 μm	3 μm

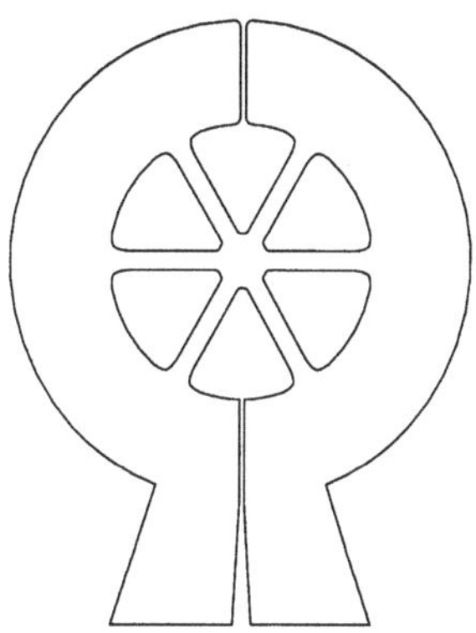

Fig. 5.6. Design of an SMA thin film microactuator consisting of three SMA bridges

Based on the actuator design valve variants 2 and 3 are developed that differ with respect to their application for gases or liquids. Fig. 5.7 schematically shows the load case of valve variants 2 and 3 in closed condition. The bridges are deformed by the spacer in the direction perpendicular to the valve seat. As for SMA beams, the force balance, Eq. 5.3, appears in closed condition. In this case, F^A denotes the force of an SMA bridge that is in balance with the closing force F_c and the pressure-induced force $F_{\Delta p}$.

In contrast to beams, a bending load, as well as a tension load occurs in the case of bridges. However, for large strain above 0.5% the bending load can be neglected in first approximation. In this case, the force of the bridges is estimated analytically by a simplified model, which is illustrated in Fig. 5.8. A more exact analysis of the mechanical properties is performed by FEM simulations.

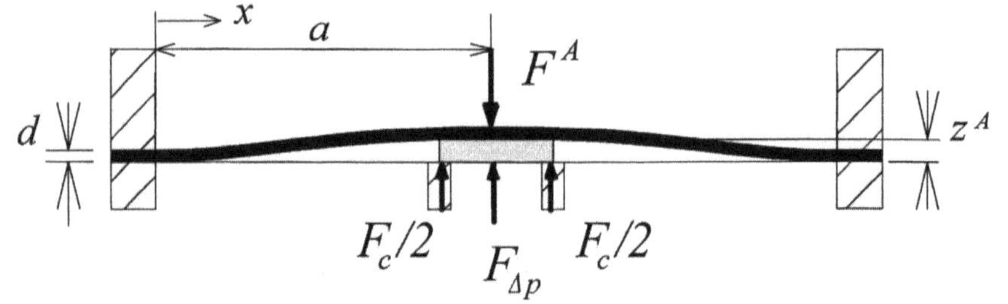

Fig. 5.7. Load case of an SMA bridge in the case of valve variants 2 and 3 in closed condition

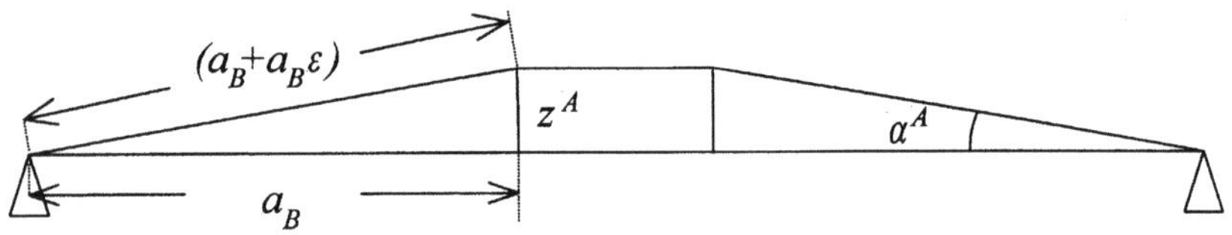

Fig. 5.8. Simplified model for analytical description of the load case in Fig. 5.7

The model assumes a point load at distance a_B from the points of suspension of the bridge. In this case, the maximum displacement of the bridge in austenitic condition z^A is:

$$z^A = \sqrt{(a_B(1+\varepsilon))^2 - (a_B)^2} \,. \tag{5.8}$$

The force F^A and the induced tensile stress σ^A are connected in the following way:

$$F^A \approx n_B b_B d \cdot \sigma^A \sin \alpha^A \,, \tag{5.9}$$

with

$$\sin \alpha^A = \frac{z^A}{a_B(1+\varepsilon)} \,. \tag{5.10}$$

Thereby, a_B, n_B and b_B denote the length, number, and width of the bridge segments. The angle of displacement of the bridges in austenitic condition is denoted by α^A. The tensile stress is determined by the tolerable load limit of the material and by requirements on the maximum work density. In the present case, the strain in the SMA microactuator is limited by the strain limit in R-phase condition of 0.8%. The resulting maximum force of the bridges and the geometry of the valve seat determine the maximum controllable pressure difference. The stroke of the SMA microactuator follows from the displacement of the bridges in austenitic and R-phase condition. Therewith the flow through the annular gap between valve seat and membrane follows as well.

Knowledge of averaged stress and strain values is not sufficient for the design of the SMA microactuator since local stress accumulations have to be taken into account. In particular, the transition regions between bridges and electrical connection paths display stress accumulations too high for designs with simple truncation. Therefore, the lateral shape of these transition regions is optimized by using the modified CAO method.

Fig. 5.9 shows an example of the simulated von Mises stress profile in austenitic condition in the optimized transition region between the SMA bridge and the corresponding electrical connection path. In the calculation, the spacer is simulated by a displacement constraint in the center part of the SMA microactuator. The resulting maximum stress is almost homogeneous in the whole transition region. In this case, the optimization criteria (O1)–(O3) are fulfilled. Based on the specification of the maximum displacement, the resulting maximum stress can be adjusted below the load limit. A displacement of 30 μm causes, e.g., a maximum von Mises stress of about 85 MPa.

The dynamics of the valve are mainly determined by the heat transfer upon heating and cooling. Thus, thermal optimization of the valve requires a good thermal coupling to its environment. For efficient cooling of the SMA microactuator, the following possibilities may be considered:

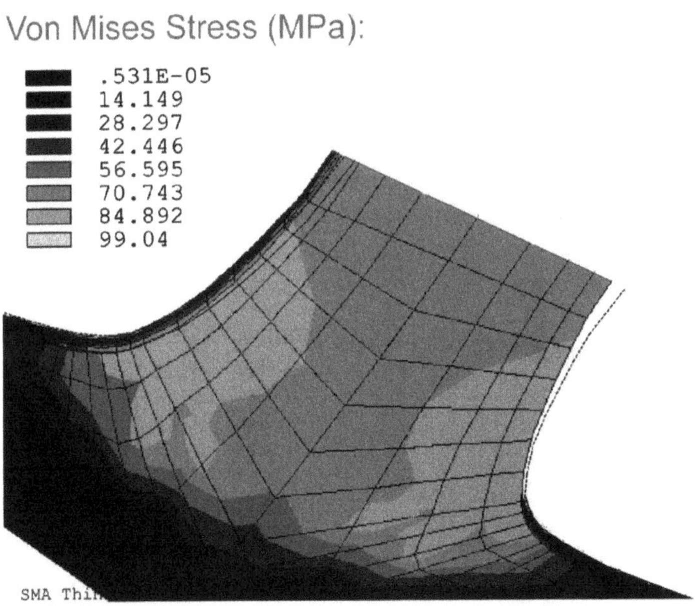

Fig. 5.9. FEM simulation of the von Mises stress profile along the surface of an SMA microactuator in austenitic condition for a spacer thickness of 30 μm. Only the transition region between the SMA bridge and electrical connection path is shown. The actuator thickness is 10 μm

- Active cooling by forced convection: The flow medium should not be used for cooling since various applications require that the temperature of the flow medium not be affected. Integration of a separate cooling circuit for active cooling seems to be very elaborate.
- Active cooling by a Peltier element: The Peltier element should preferably be integrated on the SMA microactuator in order to enable a sufficient thermal contact. Electrical isolation requires an isolation layer between the Peltier element and microactuator. In this case, however, the composite of isolation layer and Peltier layer has a large influence on the mechanical properties of the microactuator.
- Passive cooling by heat conduction: The SMA microactuator may be cooled by thermal connection to a heat sink. In this case, the SMA bridges function as thermal connectors. The heat sink may, for instance, be designed as a metallic ring, which is mounted electrically isolated above the electrical connection paths of the microactuator. This solution can be implemented quite easily and, thus, is taken into consideration in the valve design.

Fig. 5.10 shows an example of the simulated temperature profile upon electrical heating on the surface of an SMA microactuator and a metallic ring located above. In this case, the electrical supply paths are implemented separately for each bridge. Due to the symmetry of the problem, only one half of the components is considered.

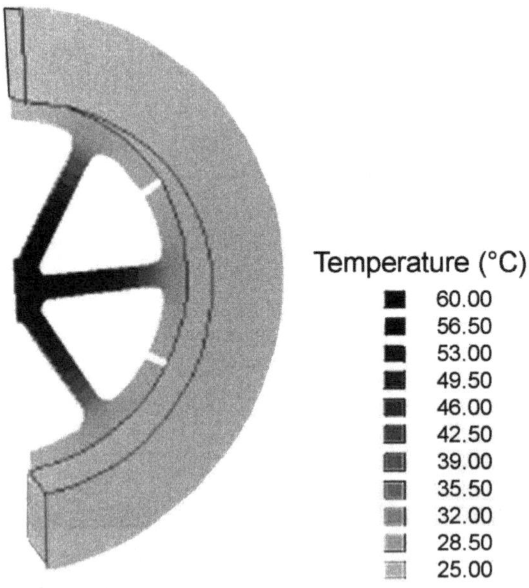

Fig. 5.10. Simulated temperature profile upon electrical heating on the surface of an SMA microactuator and a metallic ring separated by a thin electrical isolation layer

The SMA bridges show an almost homogeneous temperature profile except at the transition regions to the electrical supply paths. Thus, in particular, the regions of maximum stress and strain are heated uniformly, whereby the maximum temperature is determined by the electrical power. In this case, the optimization criteria (O4)–(O6) are fulfilled sufficiently.

The major design parameters are summarized in Table 5.6. In order to minimize swelling of the membrane during operation with a liquid, a gold layer is deposited on the lower side of the membrane of variant 3. Furthermore, the membrane of variant 3 is not micromachined to avoid a direct contact between the SMA microactuator and the liquid. In order to minimize heat transfer between the microactuator and liquid via the membrane, some prototypes of variant 3 are additionally equipped with a separation ring of a thermally isolating material such as alumina, which is mounted between microactuator and membrane. In this case, the spacer thickness is increased accordingly. Apart from these features, construction and design parameters of both variants are identical.

Table 5.6. Design parameters of SMA valves of variants 2 and 3. The length of the SMA bridge corresponds to the length $2a$ in Fig. 5.7

	Variants 2 and 3
Diameter of valve chamber	2 mm
Outer diameter of valve seat	0.3 mm
Inner diameter of valve seat	0.2 mm
Dead volume	0.45 μl
Diameter of inlet and outlet	0.2 mm
Diameter of spacer	0.7 mm
Thickness of spacer	50–100 μm
Length of SMA bridge	2 mm
Width of SMA bridge	80 μm
Number of bridges	3
Thickness of SMA actuator	20 μm
Membrane thickness	2 μm

5.3.1.3 *Variant 4 with SMA Thin Film Actuator*

Using sputtering technology, SMA thin films can be fabricated with chemical compositions, which cannot presently be obtained by cold rolling. One important example are thin films of NiTiPd. Typical thicknesses are below 20 μm. In this thickness range, the actuator design of valve variants 2 and 3 is maintained.

The thickness of the spacer, however, is adjusted to 50 μm due to safety reasons in order to limit the mechanical load. This measure proves particularly reasonable for thin films of enhanced brittleness. The thickness of the microactuators used is 10 μm. The width of the SMA bridges is adjusted to 120 μm.

Table 5.7 summarizes the major design parameters of valve variant 4. Three subvariants are fabricated, which differ in chemical composition from the SMA microactuator, as indicated in Table 5.4.

Table 5.7. Design parameters of SMA valves of variant 4

	Variant 4
Diameter of valve chamber	2 mm
Outer diameter of valve seat	0.3 mm
Inner diameter of valve seat	0.2 mm
Dead volume	0.45 μl
Diameter of inlet and outlet	0.2 mm
Diameter of spacer	0.7 mm
Thickness of spacer	50 μm
Length of SMA bridge	2 mm
Width of SMA bridge	120 μm
Number of bridges	3
Thickness of SMA actuator	10 μm
Membrane thickness	2 μm

5.3.2 Valve Fabrication

The technological approach for fabricating SMA microvalves is based on the hybrid integration of separate valve components. One objective is, thereby, to decouple critical process steps from each other to reach a high flexibility in the choice of process parameters. At the same time, maintaining the option of parallel process management in batches for the valve components does not lose track of the objective of economic production. The following section describes the fabrication of the SMA microactuators and of the valve components housing, membrane, and spacer. Finally, hybrid integration technologies will be addressed.

5.3.2.1 SMA Microactuators

After thermomechanical treatment and removal of oxide layers, the SMA materials are micromachined. For plate thicknesses on the order of 100 μm, laser cutting is particularly suitable, allowing higher aspect ratios compared to chemical etching technologies. Available etching technologies are isotropic and enable maximum aspect ratios of 2 for a double-sided application [243].

Laser cutting is performed by a (Neodym) Nd–(Ytrium-Aluminium-Garnet) YAG–solid state laser. This process may be associated with the formation of debris on the lower side of the plate caused by blown out, solidified melt. In this case, the actuators need to be deburred, which can be done wet-chemically after micromachining. In this case, a tolerance of ± 3 μm can be achieved. Due to the entry of energy during laser cutting, a heat affected zone of a few micrometers appears in the region of the cutting edge [265, 266]. However, mechanical tests do not reveal any influence on the shape memory effect.

In a freely suspended configuration, SMA thin films and foils of thicknesses in the range 10-20 μm can hardly be micromachined with sufficient precision. Therefore, a technology is required for the integration of such SMA materials on a substrate. For micromachining, wet-chemical etching is preferred compared to laser cutting, as small aspect ratios are not critical. Furthermore, chemical etching technologies can be used favorably for batch fabrication of many SMA microactuators in parallel.

As a substrate, a polished sintered alumina wafer appears suitable, which is chemically inert with respect to etching solutions applied subsequently. Integration of SMA thin films and foils may be done via a suitable sacrificial layer technology or a thin layer of adhesive. By heating the setup of SMA foil–adhesive–substrate to about 60 °C, the different thermal expansion coefficients of substrate and SMA foil lead to a completely planar layer system. This effect is supported by the planar memory shape of the SMA material. In the case of electrical connection by anisotropic conduction strips, an additional layer of Au is sputtered onto the SMA thin film.

Now, micromachining of the SMA thin film or foil is performed by optical lithography and a subsequent etching step. Thereby, either a pure chemical or an electrolytic etching technology, which has been developed in cooperation with Cranfield University (UK) [267], are used. In the case of electrolytic etching, the free metal surfaces are etched in an electrolyte in the presence of an electrical field, whereby the layer to be micromachined is connected as an anode to a voltage source. A cathode of stainless steel is inserted in coplanar orientation to the anode. The electrolyte used is a solution of methanol and concentrated sulfuric acid. For a voltage of 8 V and an ambient temperature of 20 °C, etching rates of about 10 μm/min are achieved.

Fig. 5.11(a) shows a circular arrangement of bridges fabricated by electrolytically etching a thin film of NiTiPd. The detailed view (b) clearly demonstrates the achievable structural quality. Roughness causes local peaks of field strength, which are associated with high current densities. Thus, smooth sidewalls are favored during the etching process. Furthermore, the electrical field–induced preferred orientation allows the formation of steep sidewalls. For the material of sacrificial layer, a metal layer is chosen that does not dissolve during electrolytic etching, but still can be removed selectively with respect to the SMA thin film. For residuum-free electrolytic etching, an intact sacrificial layer is required, since only in this case an electrically conductive connection exists to the regions to be removed.

For an improved adhesion of the photoresist during etching, an additional primer layer may be used below the photoresist. Alternatively, a micromachined thin metallic layer may be used as an etch mask. Due to the undirected etching effect, the width of the SMA bridges is reduced from 140 μm to 80–120 μm depending on the material thickness.

(a)

(b)

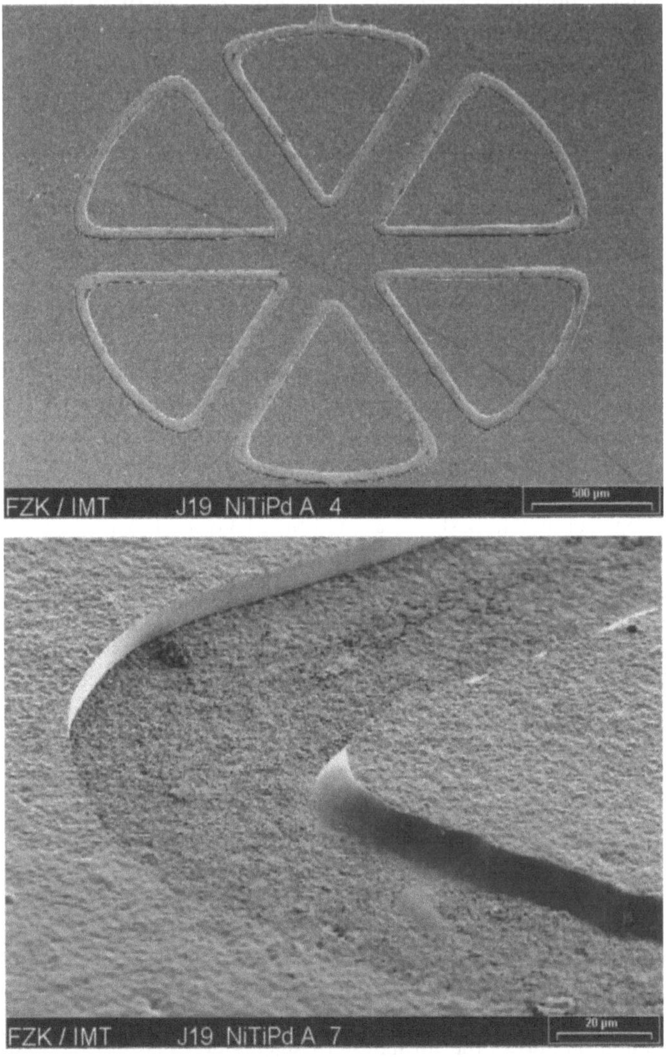

Fig. 5.11. (a): Microactuator fabricated from a sputtered NiTiPd thin film. (b): Detail of (a)

5.3.2.2 Valve Components

The upper and lower housings of the valves are fabricated out of a polymer by precision milling. In the lower housing, a round pressure chamber and valve seat are worked out. The valve seat and the top side of the housing are in-plane. Subsequently, holes are drilled for fluidic supply hoses and electrical contact pins. Batch fabrication of the valve housings by molding is possible offhand, but is not envisioned here within the scope of rapid prototyping.

The membrane is made of polyimide. Fabrication is performed by spin-coating liquid polyimide lacquer on a wafer of silicon and by subsequent curing. Beforehand, the silicon wafer is coated by a separation layer of Au for lifting-off the membrane. For pressure compensation between fluid chamber and actuation chamber, small holes may be introduced in the membrane by lithography.

Fabrication of the spacer is performed either by die cutting or lithography. In the latter case, lithography is applied for micromachining of a second layer of polyimide, which has been spun previously onto the composite of silicon wafer and thin layer of polyimide. For die cutting, foils of materials of various stiffness are used, which are fabricated either by spin-coating (e.g., polyimide) or casting using two coplanar plates (various silicone rubbers).

5.3.2.3 Hybrid Integration

Valves with SMA Thin Plate Actuator (Variant 1): The first step of valve mounting is the transfer of a polyimide membrane onto the lower valve housing. For this purpose, the lower valve housing is bonded onto the composite of silicon wafer and thin layer of polyimide using an epoxy resin. After curing, the combination of lower housing and membrane is lifted off the silicon wafer. Now, the membrane is tightly spanned above the valve seat. Residues of Au are removed by an aqueous solution of potassium iodide and iodine. Subsequently, the spacer and SMA microactuator are bonded on the membrane using fit-up aids. Finally, the upper and lower housings are interconnected by adhesive bonding.

In the case of a lithographically fabricated spacer, another mounting sequence is required, since lifting off the combination of lower housing and membrane would result in an upside down positioning of the spacer. Therefore, in this case, the membrane is lifted off in combination with the upper valve housing, after the SMA microactuator and upper valve housing have been bonded onto the membrane.

Electrical contacts are realized by inserting metallic contact pins into designated holes with separation compatible to standard IC sockets. The contact pads of the SMA microactuator and the contact pins are interconnected electrically, e.g., by wire bonding. Flexible fluidic connections are realized by adhesive bonding of standardized capillaries into designated flow channels. Fig. 5.12 shows a top view of SMA microvalves of variants 1a and 1b. For a better view of the SMA microactuator, the upper housing has been removed. The overall sizes of the housings are 20 x 11 x 3 mm^3 for variant 1a and 10 x 11 x 3 mm^3 for variant 1b.

The presented concept of using the lower valve housing simultaneously as a carrier plate for fluidic channels and electrical connections may be easily extended to the integration of additional valves or other fluidic components in order to realize fluidic microsystems. Fig. 5.13 shows, for instance, two integrated SMA microvalves of variant 1 on a common carrier plate that are interconnected to a 3/2 normally open valve.

(a) (b)

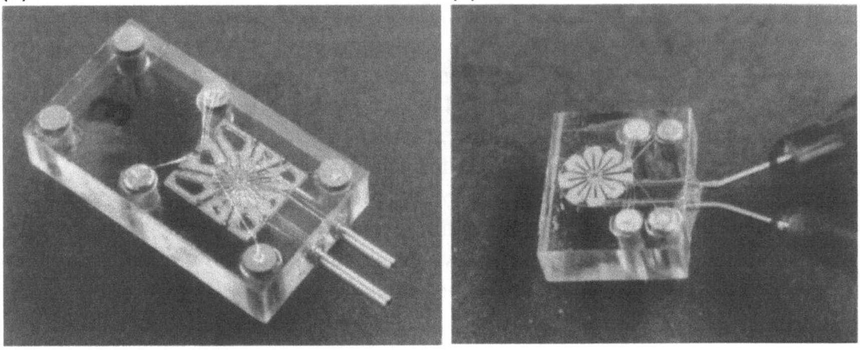

Fig. 5.12. SMA microvalves of the variants 1a (a) and 1b (b)

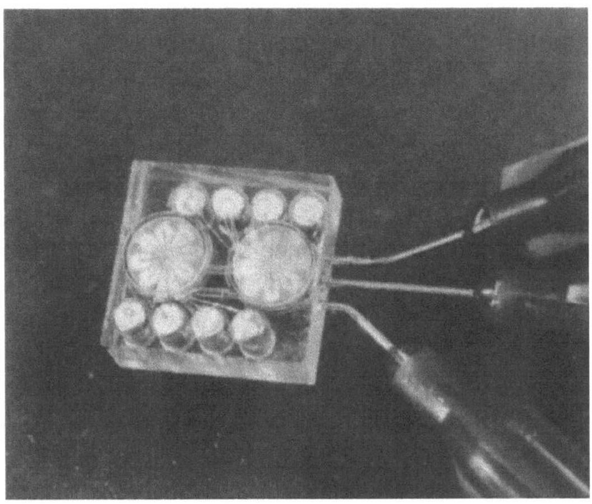

Fig. 5.13. 3/2 normally open valve consisting of two 2/2 normally open SMA microvalves of variant 1

Valves with SMA Foil and SMA Thin Film Actuator (Variants 2–4): Manual integration of ultrathin SMA microactuators on the lower valve housing appears to be very delicate. Therefore, a hybrid integration technology has been developed for the handling of highly sensitive, freely suspended microdevices. The process sequence is illustrated in Fig. 5.14.

The starting point is a composite of substrate, sacrificial layer, and SMA microactuator. By selective removal of the sacrificial layer, freely suspended microactuators are formed. These remain connected via ligaments with unstructured regions, which are still fixed to the substrate. The ligaments are designed as predetermined breaking points. As shown in Fig. 5.14, the upper housing is now bonded to the freely suspended SMA microactuator. Subsequently, the composite of upper housing and SMA microactuator is lifted off the substrate, whereby the breaking points are cut.

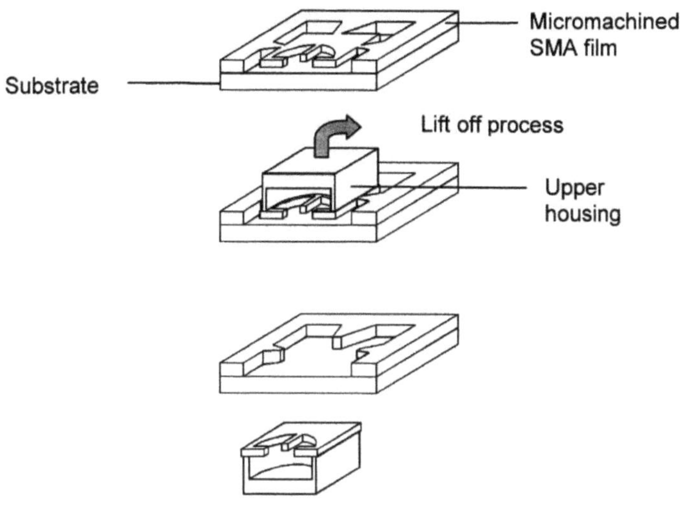

Fig. 5.14. Technology for hybrid integration of SMA microactuators

The membrane and spacer can be integrated to the lower housing in the same way as for valve variant 1. Finally, the SMA microactuator is oriented with respect to the spacer by fine positioning of the upper housing. Then, both housings are connected by adhesive bonding. In the case of a lithographically fabricated spacer, the membrane is lifted off by the composite of upper housing and SMA microactuator and then interconnected with the lower housing.

Like monolithic integration technologies described in Sect. 4.5.1, the presented hybrid integration technology is compatible with batch fabrication processes. Compared to monolithic integration the hybrid integration concept features several advantages, which may be summarized as follows:

- Separate processing of SMA microactuators allows a high flexibility during fabrication of SMA material, thermomechanical treatment, and micromachining.
- The functionality of other valve components is not affected. In particular, SMA microactuators can be combined with polymer housings of limited temperature stability.
- The problem of handling ultrasensitive microparts is shifted to the much less sensitive handling of composites consisting of valve housing and SMA microactuator.
- The use of such composites simplifies the fine positioning of the SMA microactuator with respect to the spacer and valve seat.
- By hybrid integration, modular systems can be realized in a flexible way.

Electrical and fluidic contacts can be made in the same way as valve variant 1. Fig. 5.15 shows a finished SMA microvalve of variant 4. The overall dimensions of the housing are 6 x 11 x 3 mm^3.

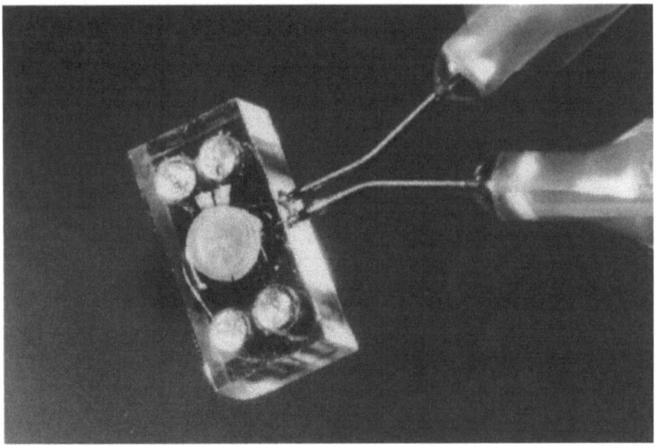

Fig. 5.15. SMA microvalve of variant 4. The outer diameter of the upper housing is 2.5 mm

Alternatively, the electrical contacts may be realized by laser welding or by using anisotropic conduction strips. Laser welding may be used in combination with a ribbon cable or a thin conducting plate, which is pressed onto the electrical connection paths of the SMA microactuator using a tension device. The application of adjusted pulses of a focused laser beam then allows selective welding of the joint. Fig. 5.16 shows, e.g., a microvalve of variant 2 with laser welded contacts.

Anisotropic conduction strips are inserted in designated openings on both sides of the valve chamber, which may be designed as part of a fluidic carrier plate. Then, the carrier plate and a cover are pressed onto a base plate with integrated conductor paths and interconnected in this configuration by adhesive bonding. By compressing the conduction strips, conductive pins pierce the flexible strip and, thus, provide an anisotropic electrical connection to the base plate.

Fig. 5.16. SMA microvalve of variant 2 with laser welded contacts

This technology is of interest particularly for modular integration of different fluidic components on a common base plate. Fig. 5.17 shows, for instance, a top and side view of an SMA microvalve of variant 3, which has been electrically connected using anisotropic conduction strips. For a better view of the SMA microactuator, the valve cover has been removed.

Fig. 5.18 shows four SMA microvalves of variant 3, which have been integrated on a common base plate in a modular way. The base plate contains integrated conductor paths and fluidic channels with ports for external interconnection. Thus, other fluidic components may also be combined arbitrarily, in addition to the microvalves, in order to realize fluidic microsystems in a flexible way. Fig. 5.19(a) shows, for instance, a microfluidic controller consisting of an SMA microactuator and a flow sensor [268]. In this case, the fluidic and electrical connections are arranged in different layers, as sketched in Fig. 5.19(b).

(a) (b)

Fig. 5.17. Top view (a) and side view (b) of an SMA microvalve of variant 3. The inner diameter of the metal ring is 2.3 mm

Fig. 5.18. SMA microvalves of variant 3 on a common base plate containing integrated conductor paths and fluidic channels

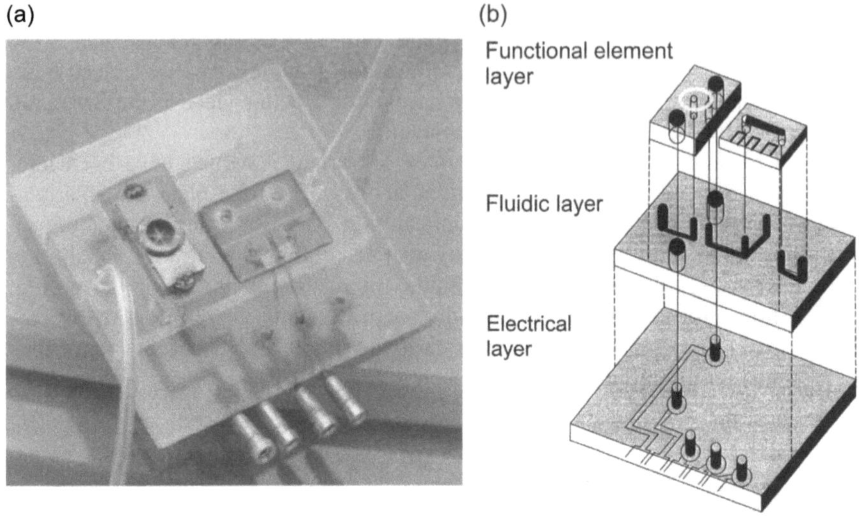

Fig. 5.19. (a) Microfluidic controller consisting of an SMA microactuator and a flow sensor, (b) exploded view of the microfluidic controller

5.3.3 Performance Characteristics

The valve properties are investigated in time-independent stationary condition, as well as a function of time using pulsed electrical current signals. Stationary equilibrium conditions are adjusted by cycling the electrical heating power step-wise with a sufficient waiting time for every data point. The minimum waiting time is automatically determined by measuring the change of resistance, which should decrease below a given tolerance limit.

By simulating the valve properties, the mechanical and thermal loads on the valve actuator are determined. For this purpose, the analytic models described in Sect. 5.3.1 and the FEM models presented in Sect. 4.2 are used. The valve specifications are determined based on the load limits given by the material properties.

5.3.3.1 Variant 1 with SMA Thin Plate Actuator

Stationary Performance: Typical characteristics of membrane stroke, gas flow, and electrical resistance are shown in Figs. 5.20–5.22. The indicated pressure differences have been determined with respect to a reference pressure of 100 kPa.

For the maximum controllable pressure difference of 120 kPa, a maximum stroke of the membrane center of 70 μm is observed in open condition. In this case, the maximum gas flow is 1,600 standard ccm. For closing the valve, an

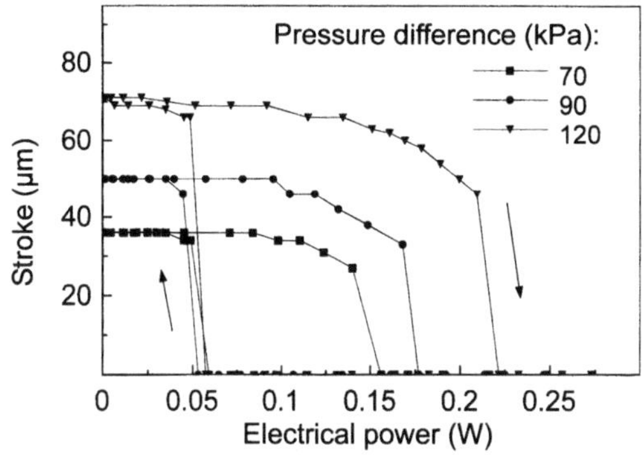

Fig. 5.20. Membrane stroke as a function of electrical heating power for an SMA microvalve of variant 1a [260]

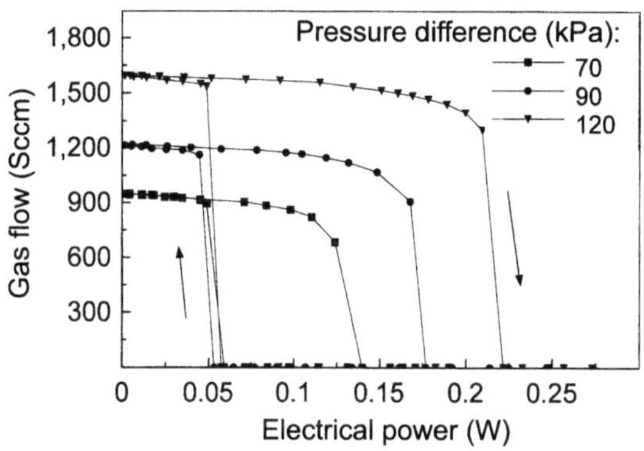

Fig. 5.21. Gas flow as a function of electrical heating power for an SMA microvalve of variant 1a [260]

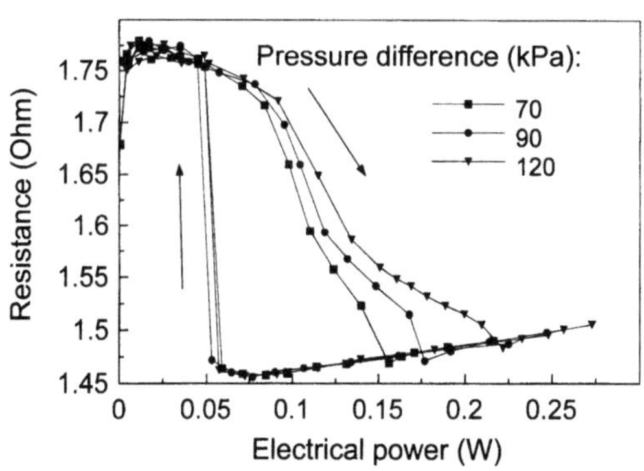

Fig. 5.22. Electrical resistance as a function of electrical heating power for an SMA microvalve of variant 1a [260]

electrical heating power below 210 mW is required. The leak rate in closed condition is below 0.1%.

The progression of resistance reflects the process of phase transformation in the SMA actuator. On average, the austenitic start temperature is exceeded at a heating power of about 80 mW. The austenitic finish temperature is reached for a heating power between 160 and 210 mW depending on the applied pressure difference. Upon cooling, the start and finish temperatures of R-phase transformation are passed through in a narrow range of heating power between 45 and 55 mW.

The characteristics show clear differences with respect to the opening and closing behavior. Upon closing the valve, the required heating power increases significantly for increasing pressure difference. The critical heating power below which the valve opens, however, is almost independent of the applied pressure difference. This performance is caused by the different cooling performance of the SMA actuator in open and closed condition. In closed condition, the SMA actuator is mainly cooled by heat conduction. If the electrical heating power is smaller than the losses by heat conduction, the temperature of the SMA actuator will decrease and the valve will open once the temperature falls below the R-phase transformation temperatures. In open condition, convective cooling also contributes. Convective cooling is caused by the gas flow and, therefore, increases with rising pressure difference. In order to close the valve, the losses of heat conduction and convection have to be compensated for by the electrical heating power. Consequently, the heating power required for closing increases with rising pressure difference. The hysteresis width observed in the characteristics can, therefore, not be attributed to the behavior of phase transformation.

An analysis of the mechanical valve behavior at maximum controllable pressure difference of 120 kPa results in the properties summarized in Table 5.8. In this case, every beam of the SMA actuator of variant 1a exerts a force F^A of about 94 mN on the valve seat. For variant 1b, F^A is 65 mN. By analytic calculations, the maximum stress in austenitic condition σ^A is estimated to be 170 MPa for variant 1a and 130 MPa for variant 1b. The corresponding strains ε^A are 0.32 and 0.24%, respectively. The maximum strain in R-phase condition ε^R is about 0.8% and, thus, reaches the given limit.

Table 5.8. Mechanical properties of microvalves of variant 1 for a pressure difference of 120 kPa

	Variant 1a	Variant 1b
σ^A	170 MPa	130 MPa
ε^A	0.32%	0.24%
ε^R	0.8%	0.8%
F^A	94 mN	65 mN
$z^A - z^R$	70 μm	80 μm

The fluidic properties summarized in Table 5.9 are determined on the basis of the mechanical properties. For a width of the valve seat of 250 μm, the critical Reynolds number Re_t is determined to be 40. Valve variant 1b shows a critical Reynolds number Re_t of only 8 due to the small width of the valve seat of 50 μm. At 120 kPa, the critical Reynolds numbers are exceeded by the actual Reynolds numbers Re by more than a factor of 100. The membrane stroke above the valve seat is estimated to about 55 μm for variant 1a and about 60 μm for variant 1b. Due to the small L/D_h ratio, it is assumed that a not fully developed flow with dominating inertial losses is present. Therefore, the flows Q can be determined by Eq. 5.1. The calculated results are somewhat higher than the observed values.

Table 5.9. Fluidic properties of microvalves of variant 1

	Variant 1a	Variant 1b
Re_t	40	8
Re	4,200	3,500
L/D_h	1	0.21
p_{in}	220 kPa	220 kPa
p_{out}	100 kPa	100 kPa
Q	2,000 standard ccm	2,900 standard ccm

The relatively large width of the valve seat of variant 1a may cause a strong adhesion between membrane and valve seat, whereby the required minimum pressure for opening inreases correspondingly. This disadvantage is avoided in valve design 1b. During the fabrication of this design, however, particular attention has to be paid to the planar surface of the valve seat in order to achieve a satisfactory sealing.

Pressure-compensated valve designs with fluidic interconnection between fluid and actuation chambers allow significantly higher maximum controllable pressure differences. Fig. 5.23 shows, for instance, gas flow characteristics of a pressure-compensated microvalve. In this case, a fluidic resistance has been inserted at the outlet to reduce the flow. In this case, the maximum controllable pressure difference is about 450 kPa.

Dynamic Performance: Opening and closing times of the valves are determined by time-resolved flow measurements. In this case, the closing time is determined by the time difference that the flow requires to reach its minimum. Accordingly, the opening time follows from the time the maximum flow is reached.

Fig. 5.24 shows opening and closing times as a function of pressure difference for various values of heating power. The closing time increases for rising pressure difference due to convective cooling, which increases simultaneously. For a pressure difference of 120 kPa and an electrical heating power of 450 mW, a closing time of 1.2 s is reached. This time reduces to 0.5 s by decreasing the pressure difference to 40 kPa.

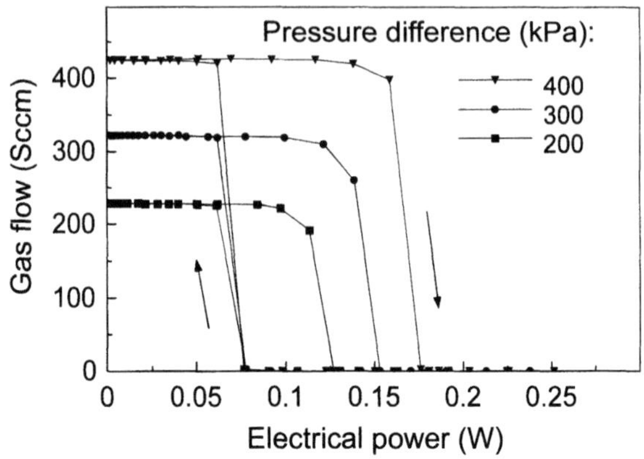

Fig. 5.23. Gas flow as a function of electrical heating power for a pressure-compensated SMA microvalve of variant 1a [260]

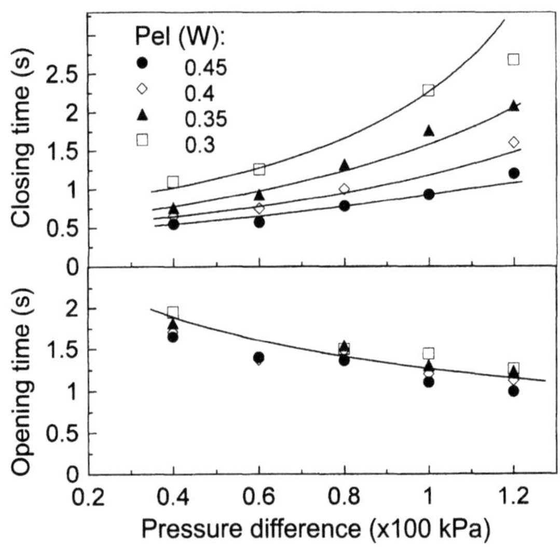

Fig. 5.24. Opening and closing times as a function of the pressure difference and heating power [260]

The solid lines are calculated closing times, which can be estimated from the balance of power during valve operation:

$$\tau_s = \frac{m \cdot (c_p \cdot \Delta T + \Delta H) + U}{P_{el} - P_{conv} - P_{cond}}. \tag{5.11}$$

The parameters τ_s, ΔT, U, P_{el}, P_{conv}, and P_{cond} denote the closing time, the difference between room temperature and transformation temperature, the upstroke work, the electrical heating power, and the convective and conductive heat losses, respectively. The heat losses can be determined from the stationary flow characteristics, since the heating power required for closing corresponds to the sum of P_{conv} and P_{cond} and the minimum heating power of opening is given by P_{cond}.

The opening times are less strongly affected by the pressure difference, since heat convection only partially contributes to the cooling of the SMA actuator. The opening times decrease for increasing pressure difference and vary between 1 and 2 s. By setting the electrical heating power in Eq. 5.11 equal to zero and fitting the convective contribution of cooling to the measurement results by a factor k_0, the following relationship for the opening time τ_o is obtained:

$$\tau_o = \frac{m \cdot (c_p \cdot \Delta T + \Delta H) + U}{-k_0 P_{conv} - P_{cond}}, \tag{5.12}$$

which is shown in Fig. 5.24 as a solid line. In the present case, the fraction of convective cooling amounts to 80% ($k_0 = 0.8$).

5.3.3.2 Variant 2 with SMA Foil Actuator

Stationary Performance: Fig. 5.25 shows typical characteristics of gas flow and electrical resistance of a microvalve of variant 2. The valve membrane contains holes for pressure compensation between valve chamber and actuation chamber. The spacer consists of silicone of thickness 60 µm. The cooling ring consists of copper. The indicated pressure differences have been determined with respect to a reference pressure of 100 kPa. The maximum controllable pressure difference is about 500 kPa. At this pressure difference, the valve allows a maximum flow of 470 standard ccm. For increasing heating power, the flow decreases until a critical pressure-dependent heating power is reached, above which the valve closes. The heating power required for closing strongly depends on the pressure difference. For 500 kPa, it reaches about 67 mW. The leak rate in closed condition is below 1 standard ccm.

For pressure differences below 200 kPa, the characteristics of gas flow and electrical resistance display a narrow hysteresis. In this case, they reflect the typical behavior of R-phase transformation. For higher pressure differences, the hysteresis broadens increasingly, as convective cooling causes enhanced power consumption. The strong increase of hysteresis above 400 kPa is additionally caused by SIM formation, which occurs for strains above 0.6% [140].

Compared to microvalves driven by a NiTi thin plate actuator of 100 µm thickness, the valve variant presented allows a significantly smaller maximum flow. Besides the smaller dimensions of the valve seat, this behavior mainly results from the relatively small membrane stroke, which reaches a maximum of about 18 µm due to R-phase transformation.

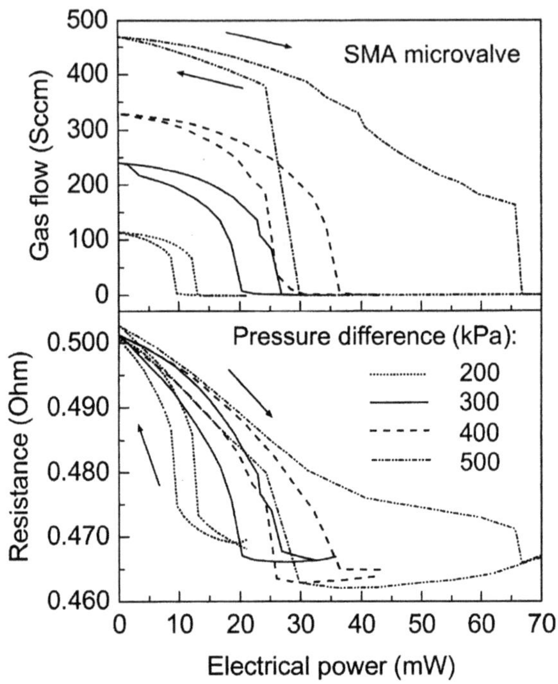

Fig. 5.25. Gas flow and electrical resistance as a function of electrical heating power for a pressure-compensated SMA microvalve of variant 2 [269]

In contrast, the maximum controllable pressure differences reach comparable values. This behavior is attributed to the better use of the shape memory effect in bridge structures compared to beam structures, since high stress and strain values not only occur in the surface layers but in the whole volume due to tensile loading.

An analysis of the mechanical valve behavior at a pressure difference of 500 kPa results in the properties summarized in Table 5.10. In this case, the vertical displacements in open and closed conditions z^R and z^A give rise to a stroke of the SMA actuator of about 18 μm. In closed condition, every SMA bridge exerts a force F^A of about 80 mN on the valve seat. The maximum closing force of the valve cannot be determined by a simple estimate based on pressure difference between inlet and outlet and area of valve seat, since the pressure distribution below the membrane strongly deviates from a constant distribution. From the maximum values of force and stroke of the SMA actuator the maximum work density is calculated to be about 50 mJ/g. The maximum tensile stress in the SMA actuator is estimated analytically to be 255 MPa. The maximum strain in austenitic and R-phase conditions, ε^A and ε^R, is about 0.45 and 0.8%, respectively. Thus, the given strain limits are reached for both phase states. For comparison, Table 5.10 also contains the maximum von Mises stress and effective strain in the SMA actuator determined by FEM analysis.

Table 5.10. Mechanical properties of a microvalve of variant 2 for a pressure difference of 500 kPa. The analytically determined maximum tensile stress and strain are denoted by $\sigma^{A,R}$ and $\varepsilon^{A,R}$, the maximum von Mises stress and effective strain determined by FEM analysis by $\bar{\sigma}^{A,R}$ und $\bar{\varepsilon}^{A,R}$

Analytic results		FEM results	
σ^A	255 MPa	$\bar{\sigma}^A$	290 MPa
ε^A	0.45%	$\bar{\varepsilon}^A$	0.53%
σ^R	195 MPa	$\bar{\sigma}^R$	130 MPa
ε^R	0.8%	$\bar{\varepsilon}^R$	0.56%
F^A	80 mN		
$z^A - z^R$	18 μm		

For austenitic condition, both models show a good agreement due to a homogeneous stress distribution in the SMA bridges. The maximum von Mises stress determined by FEM analysis deviates from the analytical value by about 12%. Therefore, elaborate FEM simulations are not required for determining maximum tensile stress in closed condition. However, in R-phase condition, a significant deviation from the analytical result is observed. This is due to the non-linear progression of the stress-strain characteristic, which is not taken into account by using an effective E-modulus. Therefore, the analytical calculation overestimates the tensile stress in R-phase condition by about 30%. Furthermore, FEM simulations reveal an inhomogeneous stress distribution in R-phase condition characterized by a maximum stress close to the edge of the spacer.

The fluidic properties summarized in Table 5.11 follow from the mechanical properties for a pressure difference of 500 kPa. For a valve seat width of 50 μm, the critical Reynolds number, Re_t, is about 25. In contrast, the Reynolds number, Re, calculated for 500 kPa amounts to about 2,980. For a stroke of 18 μm, the L/D_h ratio is 0.7. Thus, it is assumed that the flow is not fully developed and that inertial losses dominate. The maximum flow Q is calculated by Eq. 5.1 to 480 standard ccm.

Table 5.11. Fluidic properties of a microvalve of variant 2

Re_t	25
Re	2,980
L/D_h	0.7
p_{in}	600 kPa
p_{out}	100 kPa
Q	480 standard ccm

Dynamic Performance: The dynamic performance of the valves is investigated by time-resolved gas flow measurements. Gas flows are determined by the differential pressure principle using two pressure sensors. Fig. 5.26 shows the behavior of closing times as a function of electrical heating power. The closing times decrease in inverse proportion to the heating power, as expected for adiabatic heating using short pulses of heating power. Within the investigated range of heating power below 90 mW the minimum closing time is about 25 ms. The maximum work density of 50 mJ/g and the minimum closing time correspond to a maximum power density of the SMA microactuator of 2 W/g.

The time-resolved progression of gas flow upon closing and opening is shown in Fig. 5.27. In this case, a closing time of about 20 ms is obtained for a heating power of 220 mW. Interruption of heating power after valve closing without the supplying of extra heat gives rise to a power-independent opening time, τ_o, of about 50 ms. Consequently, the maximum operation frequency for complete opening and closing is about 15 Hz.

For simulation of spatially and time-resolved temperature distributions, the distributions of heating power must be determined first. As explained in Sect. 4.2.5, these distributions follow from electrical FEM simulations of corresponding distributions of electrical current. In the investigated temperature range below 100 °C, the mechanisms of heat conduction and convection are mainly responsible for cooling of the SMA actuator. To allow for a realistic description of heat conduction, the bonding layer and metal ring are taken into account in the simulation. The effect of convective cooling of the SMA actuator is simulated by using an effective heat-transfer coefficient $\overline{K}$. This coefficient is determined by fitting simulated temperature profiles to corresponding experimental profiles of an SMA test sample. For this purpose, different cases of free and forced convection are investigated using an infrared camera. Finally, the latent heat of phase transformation is taken into account.

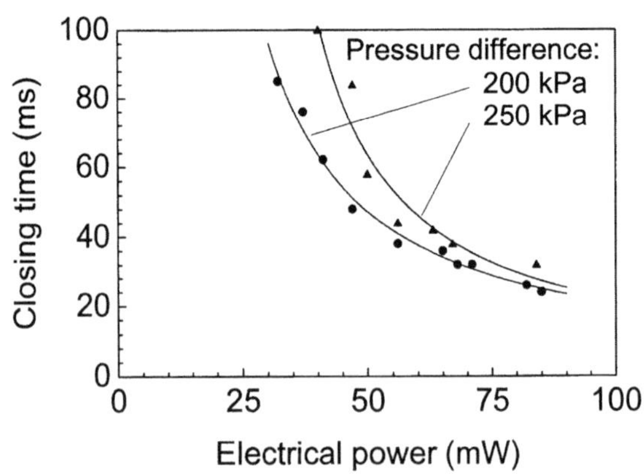

Fig. 5.26. Closing times as a function of electrical heating power for a microvalve of variant 2 [257]

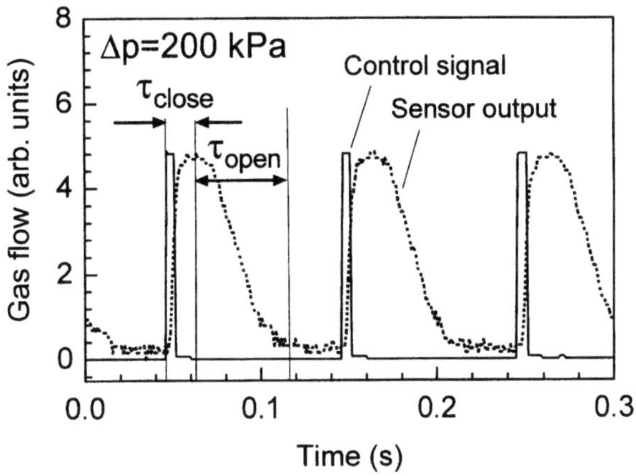

Fig. 5.27. Time-resolved gas flow upon closing and opening of a microvalve of variant 2 [268]

Table 5.12 contains a survey of the simulation parameters used. Within the temperature range of 20 to 60 °C, the parameters are treated as constants.

Table 5.12. Simulation parameters for the determination of the temperature profile in a microvalve of variant 2

	SMA actuator	Bonding layer	Cover
Thermal conductivity λ ($Wm^{-1}K^{-1}$)	18	0.02	380
Electrical conductivity ρ^{-1} ($\Omega^{-1}m^{-1}$)	$1.2 \cdot 10^6$	0.5	$50 \cdot 10^6$
Specific heat capacity c_p ($Jkg^{-1}K^{-1}$)	450	200	385
Mass density ρ (kgm^{-3})	$6.5 \cdot 10^3$	$4 \cdot 10^3$	$8.8 \cdot 10^3$
Thermal expansion coefficient α (K^{-1})	10^{-5}	$6 \cdot 10^{-5}$	$1.7 \cdot 10^{-5}$
Heat transfer coefficient $\overline{K}$		70 $Wm^{-2}K^{-1}$	
Integral latent heat of phase transformation ΔH (R-phase transformation)		4,500 Jkg^{-1}	
Environment temperature T_E		25 °C	

Fig. 5.28 shows a series of temperature profiles along the top surfaces of an SMA microactuator and a metallic ring for a heating power of 85 mW. Due to symmetry, only one half of the valve is considered. The corresponding heating characteristic in the center of the SMA microactuator is shown in Fig. 5.29.

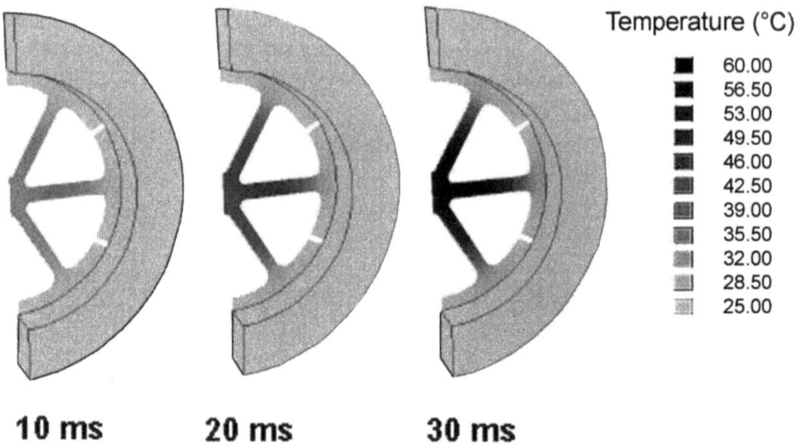

Fig. 5.28. Simulated temperature profiles along the top surfaces of the SMA microactuator and metallic ring for different heating times as indicated. The heating power is 85 mW [269]

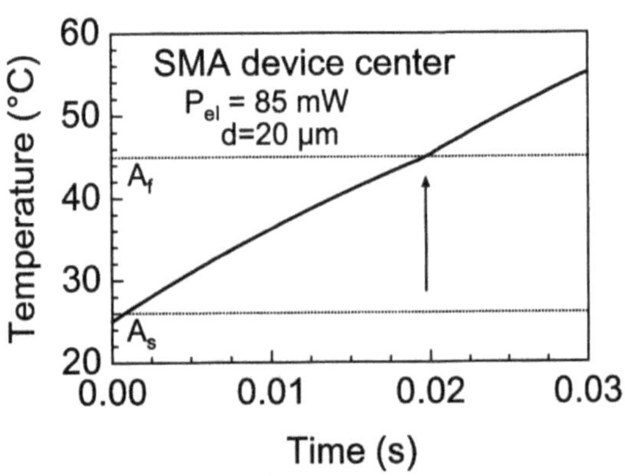

Fig. 5.29. Heating characteristic in the center of the SMA microactuator [269]

After a heating time of about 20 ms, the maximum temperature reaches the A_f temperature of 45 °C. After about 25 ms, more than 95% of the SMA bridges has already transformed to austenite. Thus, different regions of the bridges contribute almost simultaneously to the closing of the valve.

By switching the heating power off after a heating time of 30 ms, the sequence of temperature distributions shown in Fig. 5.30 is obtained. The corresponding cooling characteristic in the center of the SMA microactuator is shown in Fig. 5.31.

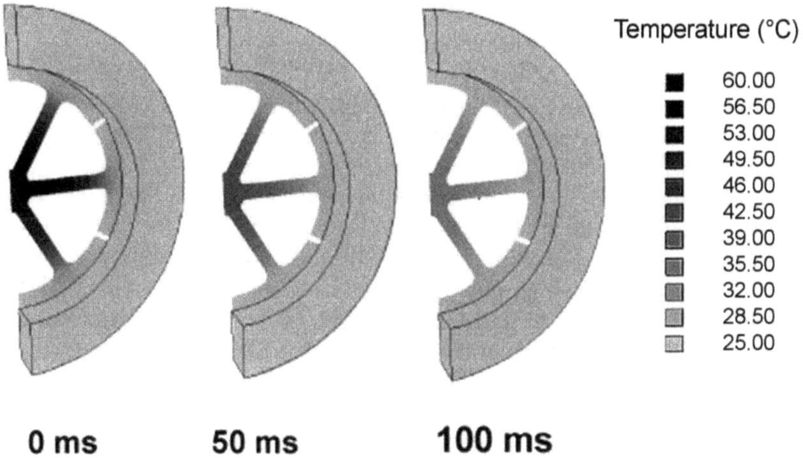

Fig. 5.30. Simulated temperature distributions along the top surfaces of the SMA microactuator and metal ring for different cooling times as indicated [269]

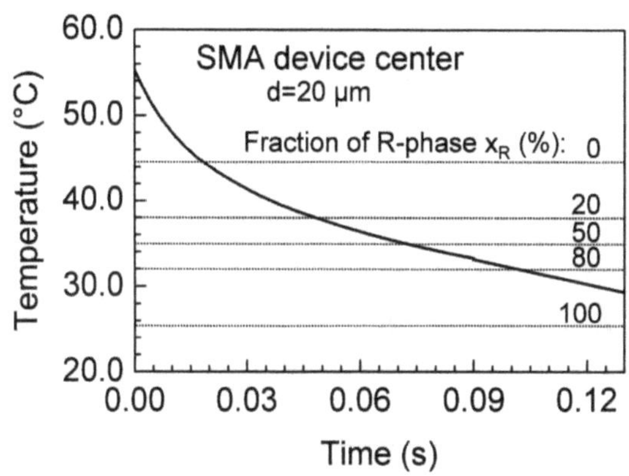

Fig. 5.31. Cooling characteristic in the center of the SMA microactuator [269]

After a cooling time of about 100 ms, the maximum temperature has decreased below 32 °C and, thus, more than 80% of austenite has transformed back to R-phase. In the regions close to the bond pads, the phase transformation is completed by more than 90%.

The simulated heat transfer times agree very well with the measured closing and opening times of the valves. Consequently, the simulation model allows the study of the possible optimizations of the dynamic performance. One possibility for improvement is active cooling by forced air convection. Fig. 5.32 shows, for instance, simulated time-resolved characteristics of maximum temperatures occurring upon heating and cooling in the SMA microactuator for different heat transfer coefficients. In this case, the heating time and power are kept at 30 ms and 85 mW, respectively, as in the previous simulations. By increasing the heat transfer coefficient by a factor of 2, the cooling time decreases by about 12%.

A considerable improvement in valve dynamics is achieved by using SMA materials with higher phase transformation temperatures. Fig. 5.33 shows simulated characteristics of maximum temperatures upon heating and cooling for SMA microactuators with various phase transformation temperatures M_f / A_f of 80/120 °C, 130/170 °C, and 180/220 °C. The separation between start and finish temperatures has been assumed to be 20 K in all cases. For comparison, the characteristic of valve variant 2 is also shown in Fig. 5.33, corresponding to the case $R_f / A_f = 25/45$ °C. In each case, the heating power has been adjusted to heat about 95% of the SMA bridges above the A_f temperature within 30 ms. The arrows indicate the cooling times that the maximum temperature in the SMA actuator requires to decrease below the M_f temperature. These times decrease strongly to about 12 ms for M_f=180 °C. In this case, a maximum operation frequency of the valve of about 40 Hz could be achieved. The NiTiPd material system allows, for instance, adjustment of the A_f temperature up to A_f =540 °C [195]. For the presented valve design, this would lead to a cooling time and maximum frequency of about 4 ms and 125 Hz, respectively.

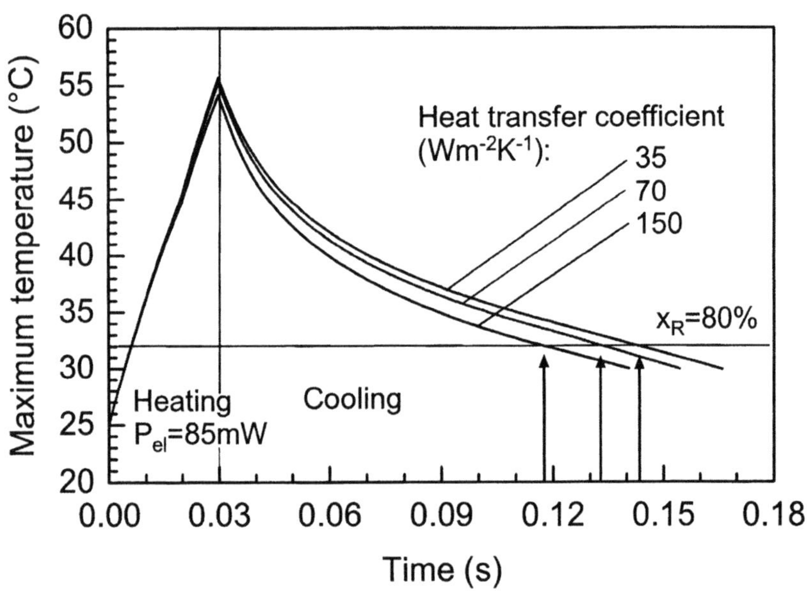

Fig. 5.32. Simulation of time-resolved maximum temperature in the SMA microactuator for different heat transfer coefficients $\overline{K}$. The temperature-dependent fraction of R-phase is denoted by x_R. The characteristic simulated for $\overline{K}$ = 70 Wm^{-2}K^{-1} corresponds to valve variant 2

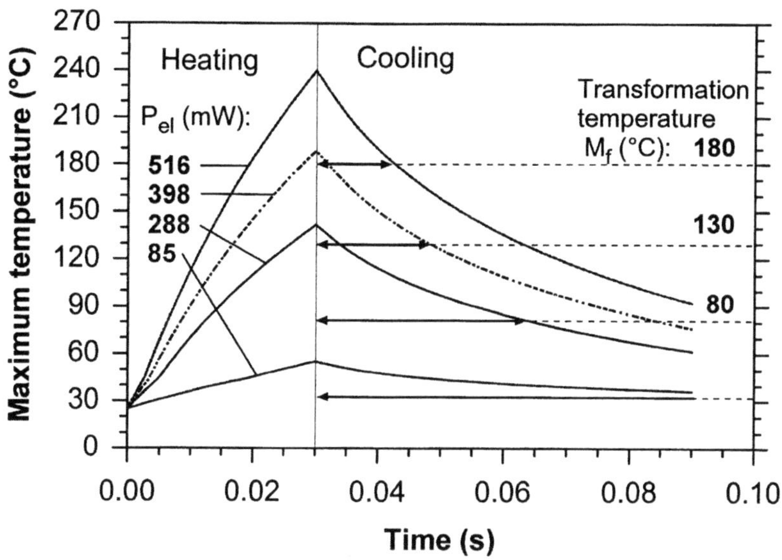

Fig. 5.33. Simulation of time-resolved maximum temperature in the SMA microactuator for SMA microactuators with various phase transformation temperatures, as represented by the martensite finish temperature M_f. The characteristic simulated for a heating power of 85 mW corresponds to valve variant 2 [257]

5.3.3.3 Variant 3 with SMA Foil Actuator

For operation with a liquid, the microactuator and liquid are kept separated to prevent a direct heat transfer. Therefore, the membrane in variant 3 microvalves is not micromachined. Consequently, the pressure difference between fluid chamber and actuation chamber is not compensated. As with variant 2, the spacer thickness of variant 3 is set to 60 µm. Fig. 5.34 shows the typical flow characteristics of water and the corresponding electrical resistance characteristics for stationary operation. In this case, the flow is determined by the pressure difference in a capillary connected in series with the valve.

As the pressure difference between fluid chamber and actuation chamber is not compensated, maximum controllable pressure differences are reduced. In the case of valve variant 3, a maximum pressure difference of about 260 kPa is achieved. At this pressure difference, a maximum flow of 13 ml/min is observed in open condition. The heating power required for closing the valve is about 100 mW. The mechanical properties of variant 3 correspond to the properties of variant 2, listed in Table 5.10. This can be expected, as with an identical valve setup, the maximum controllable pressure difference is applied in both cases.

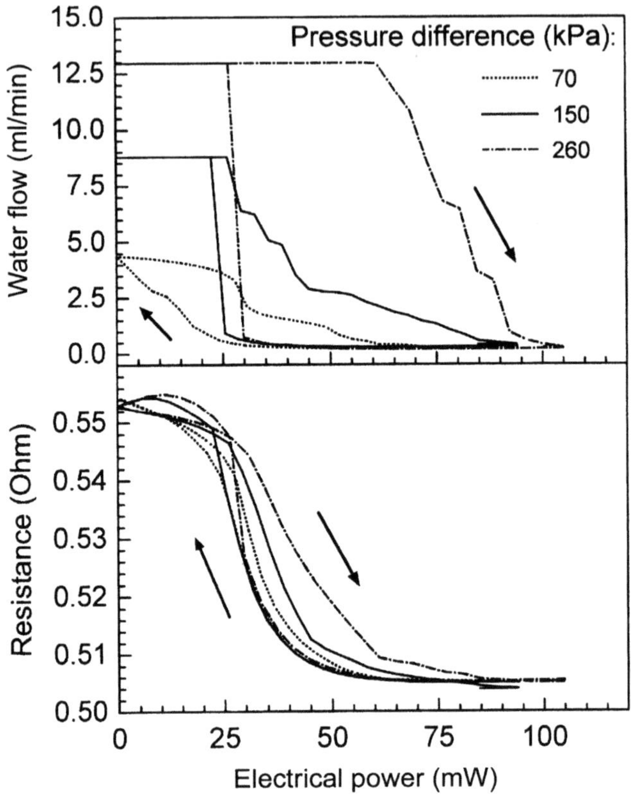

Fig. 5.34. Water flow and electrical resistance as a function of electrical power for an SMA microvalve of variant 3. The pressure differences between inlet and outlet of the microvalve are indicated

Valve variants 2 and 3 differ with respect to their thermal and fluidic properties. Thermal coupling between SMA microactuator and liquid can be much stronger than in the case of gases, which may affect the required heating power for valve operation. These effects are minimized by thermally isolating the SMA microactuator with respect to the membrane and lower housing. The characteristics shown in Fig. 5.34 are achieved, for instance, by mounting a thermal separation ring of alumina between SMA microactuator and membrane.

The fluidic properties are determined by the considerably higher viscosity of water. Also, in the present case it is assumed that the flow is not fully developed and that inertial losses dominate, since the L/D_h ratio is only 0.7 and the critical Reynolds number Re_t of about 25 is largely exceeded. The maximum flow at 260 kPa pressure difference is estimated using Eq. 5.2 to be about 12 ml/min, which agrees well with the experimental result. The fluidic properties determined for a pressure difference of 260 kPa are summarized in Table 5.13.

Table 5.13. Fluidic properties of a microvalve of variant 3

Re_t	25
Re	$4.4 \cdot 10^5$
L/D_h	0.7
p_{in}	401 kPa
p_{out}	141 kPa
Q	12 ml/min

The dynamic valve properties depend particularly on the degree of thermal coupling between SMA microactuator and liquid. Fig. 5.35 shows a time-resolved flow characteristic of a valve of variant 3 for operation with water at 200 kPa pressure difference. In this case, the time constants for complete opening and closing are 35 and 50 ms, respectively. Compared to operation with nitrogen gas, a small increase of the time constant for closing is observed, which is caused by some heat loss during electrical heating of the SMA microactuator due to enhanced heat convection. The time constant for opening is almost unaffected by the fluid used. The resulting maximum operation frequency for complete opening and closing is about 11 Hz. These results have been obtained for valves equipped with a separation ring to thermally decouple the microactuator and liquid. Valve variants without separation rings exhibit a considerably larger time constant for closing and require greater heating power. Furthermore, an increased input of heating power leads to additional heating of neighboring valve components close to the SMA microactuator, causing an increase in the time constant for valve opening.

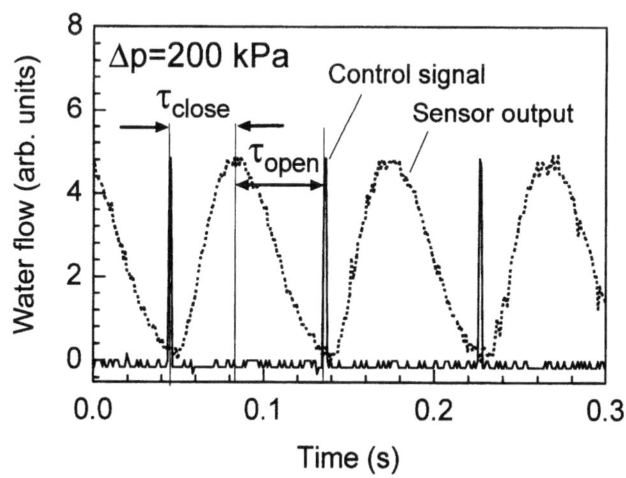

Fig. 5.35. Time-resolved water flow upon closing and opening of a microvalve of variant 3. The pulse frequency is 10 Hz [268]

5.3.3.4 Variant 4 with SMA Thin Film Actuator

Fig. 5.36 shows typical characteristics of gas flow and electrical resistance of a pressure-compensated microvalve with a thin film microactuator of NiTi operated in stationary condition. In this case, the spacer consists of silicone. The indicated pressure differences have been determined with respect to a reference pressure of 100 kPa. The maximum controllable pressure difference is about 210 kPa. In the absence of electrical heating power, the valve exhibits a maximum flow of about 180 standard ccm. For increasing heating power, the flow decreases and eventually drops to zero above a pressure-dependent critical power. In the present case, the critical power is below 30 mW. The leakrate in closed condition is below 1 standard ccm.

The characteristics in Fig. 5.36 display the typical behavior of R-phase transformation. Therefore, it can be concluded that convective cooling has no significant effect on the closing behavior for the gas flows observed.

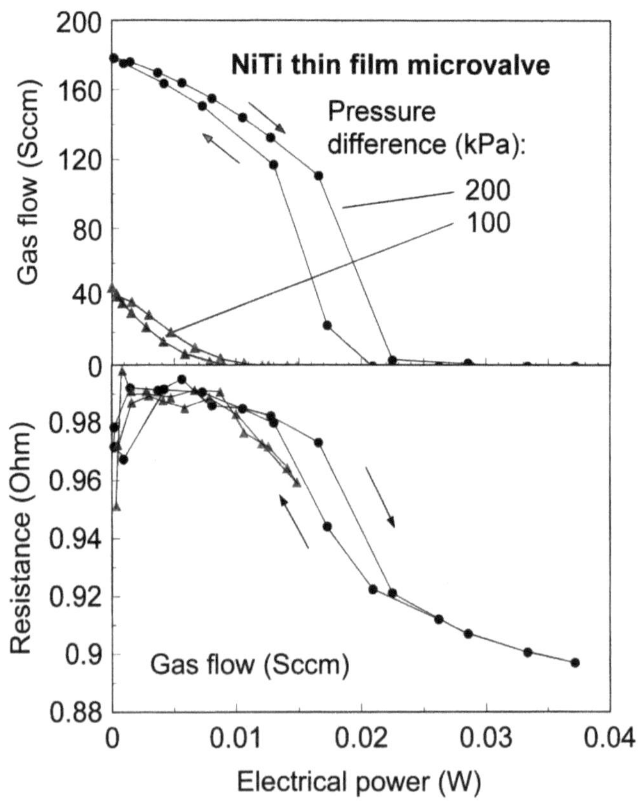

Fig. 5.36. Gas flow and electrical resistance as a function of electrical heating power for a pressure-compensated SMA microvalve of variant 4 [199]

The mechanical properties are summarized in Table 5.14 for a pressure difference of 200 kPa. The maximum tensile stress in the SMA actuator σ^A is determined analytically to 165 MPa. This value corresponds well with the maximum von Mises stress in the SMA bridges of 180 MPa determined by FEM analysis. The resulting force of one SMA bridge, F^A, is about 30 mN. As for valve variants 2 and 3, the maximum closing force of the valve cannot be determined by a simple estimate based on the pressure difference between the inlet and outlet and valve seat area, since the pressure distribution below the membrane strongly deviates from a constant distribution. The strain in austenitic condition is about 0.28%. The effective E-modulus in R-phase condition is estimated to be about 30 GPa. The stroke of the SMA actuator $z^A - z^R$ is about 12 μm.

Table 5.14. Mechanical properties of a microvalve of variant 4 with thin film microactuator of NiTi for a pressure difference of 200 kPa

σ^A	165 MPa
ε^A	0.28%
σ^R	133 MPa
ε^R	0.45%
F^A	30 mN
$z^A - z^R$	12 μm

The corresponding fluidic properties are listed in Table 5.15. The critical Reynolds number Re_t is about 38 and is largely exceeded by the calculated Reynolds number Re. For a stroke of the SMA microactuator of 12 μm, the L/D_h ratio is 1. Thus, also in the present case it is assumed that the flow is not fully developed and that inertial losses dominate. Based on the stroke of 12 μm, a maximum flow of 200 standard ccm is determined analytically.

Table 5.15. Fluidic properties of a microvalve of variant 4 with a thin film microactuator of NiTi

Re_t	38
Re	1,270
L/D_h	1
p_{in}	300 kPa
p_{out}	100 kPa
Q	200 standard ccm

Fig. 5.36 shows gas flow characteristics of a microvalve with a thin film microactuator of NiTiCu in stationary operation mode. Due to pressure compensation, the maximum controllable pressure difference is about 300 kPa. At this pressure difference, the valve exhibits a maximum flow of about 360 standard ccm. The power required for closing increases strongly with increasing pressure difference. At 300 kPa, it reaches about 100 mW.

Due to the use of a martensitic phase transformation, larger strains and thus larger membrane strokes are possible in open condition. The resulting higher gas flow in open condition is associated with a higher convective cooling and, consequently, with a higher power requirement for closing the valve. As a result, pronounced hysteresis characteristics occur particularly at higher pressure differences. At small pressure differences of 100 kPa and below, the characteristics reflect the intrinsic hysteresis of the martensitic transformation.

The mechanical properties determined for the maximum controllable pressure difference of 300 kPa are summarized in Table 5.16. Due to the higher E-modulus of NiTiCu compared with NiTi in austenitic condition, a higher force can be generated upon closing the valve. Consequently, microvalves with microactuators of NiTiCu are able to control higher pressure differences for a given spacer thickness.

Also, in the present case, it is assumed that the flow is not fully developed and that inertial losses dominate, since the L/D_h ratio is only 0.7 and the critical Reynolds number Re_t of about 26 is greatly exceeded. The fluidic properties determined for 300 kPa are listed in Table 5.17.

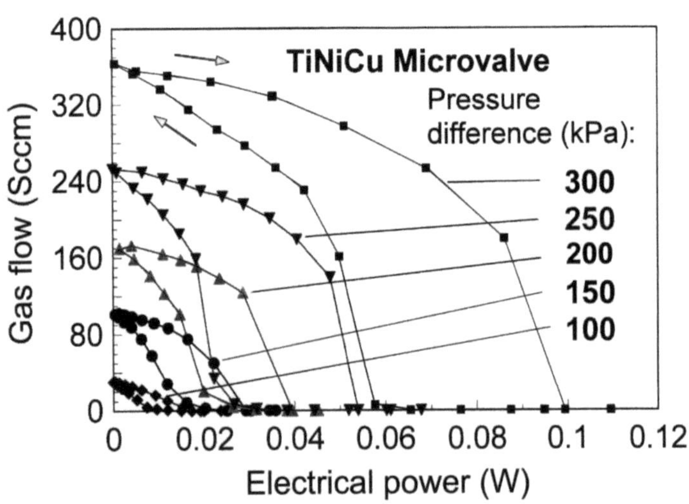

Fig. 5.37. Gas flow as a function of electrical heating power for a pressure-compensated SMA microvalve of variant 4 with a thin film microactuator of NiTiCu [259]

Table 5.16. Mechanical properties of a microvalve of variant 4 with a thin film microactuator of NiTiCu for a pressure difference of 300 kPa

σ^A	290 MPa
ε^A	0.28%
σ^M	210 MPa
ε^M	0.51%
F^A	50 mN
$z^A - z^M$	18 μm

Table 5.17. Fluidic properties of a microvalve of variant 4 with thin film microactuator of NiTiCu

Re_t	26
Re	2,280
L/D_h	0.7
p_{in}	400 kPa
p_{out}	100 kPa
Q	370 standard ccm

Fig. 5.38 shows stationary characteristics of gas flow and electrical resistance of a microvalve with a thin film microactuator of NiTiPd. In other respects, the valve setup corresponds to the previous microvalves. The maximum controllable pressure difference is about 260 kPa, which is somewhat lower than the result obtained for NiTiCu. Compared to NiTi, the maximum controllable pressure difference is considerably higher due to the higher E-modulus of NiTiPd in austenitic condition. For 250 kPa, the maximum flow is about 360 standard ccm. The power required for closing the valve strongly depends on the pressure difference and reaches about 210 mW for 250 kPa. The leakrate in closed condition is again below 1 standard ccm.

Compared to microvalves with thin film microactuators of NiTiCu, a considerably higher power is required for closing. Thereby, the gas flows are similar, so that a comparable convective cooling can be assumed. Thus, solely the higher phase transformation temperatures of the NiTiPd alloy are responsible for the higher heating power, see Table 5.4. Due to this reason, the minimum power required to maintain the closed condition also displays a relatively high value of about 50 mW. The intrinsic hysteresis of the martensitic transformation is only observed at small pressure differences below 100 kPa, as for NiTiCu microactuators.

The mechanical and fluidic properties determined for the maximum controllable pressure difference of 260 kPa correspond to the properties of microvalves with thin film microactuators of NiTiCu, listed in Tables 5.16 and 5.17.

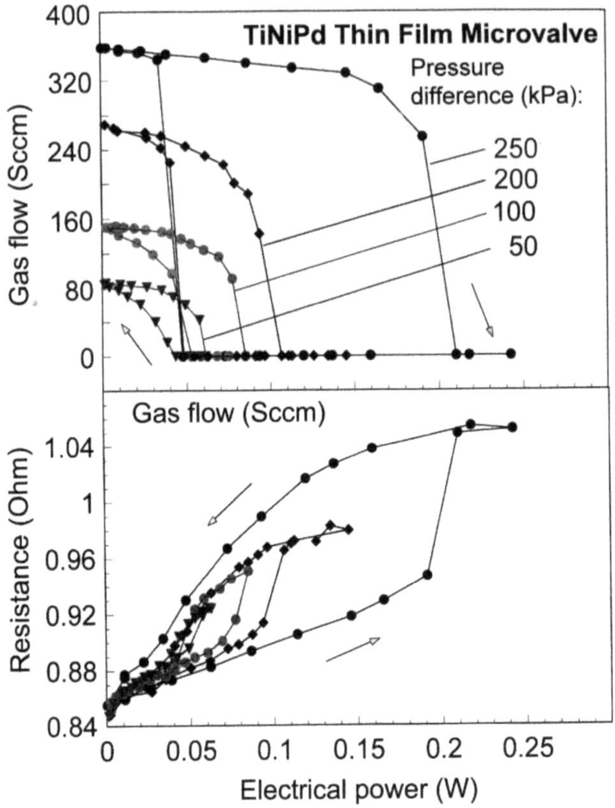

Fig. 5.38. Gas flow and electrical resistance as a function of electrical heating power for a pressure-compensated SMA microvalve of variant 4 with a thin film microactuator of Ni-TiPd [199]

5.3.4 Conclusions

Micromachined SMA actuators with a stress-optimized design are used for the operation of SMA microvalves. In the case of beams elements in valve variant 1, stress optimization gives rise to a homogeneous stress profile along the top and bottom surfaces. For SMA bridges used in valve variants 2–4 an even more homogeneous stress profile is achieved extending through the whole volume due to tensile loading. The stress in the SMA microactuators is adjusted to generate a work density as high as possible without exceeding material-dependent limits. Thus, the shape memory effect can be used in an optimum way. Furthermore, the yield of hybrid integration of SMA actuators is improved and the lifetime of the valves is extended. As a major result, fabrication and operation of NiTiPd microvalves could be demonstrated for the first time.

By using an R-phase transformation in microactuators of NiTi, the work density of the valves is limited by the maximum strain in R-phase condition of 0.8%. In this case, the microactuators achieve a maximum work density of 50 mJ/g and a power density of 2 W/g. These specifications are already comparable to macroscopic gas turbines, see Fig. 2.2 in Section 2.1.

Fabrication of the SMA microvalves is performed by hybrid integration of a few valve components (housing, membrane, spacer, SMA microactuator). A specially developed technology for hybrid integration of micromachined foils and thin films allows a trouble-free handling and precise mounting. All technology steps, including fabrication of valve components and hybrid integration, can be performed in parallel. In the present case, the SMA actuator of valve variant 1 is micromachined by laser cutting, however, which appears to be advantageous for aspect ratios larger than 1. SMA foils of 20 μm thickness and thin films are micromachined by optical lithography and chemical etching, as requirements on aspect ratios are not critical.

Table 5.18 comprises the achieved specifications of a pressure-compensated SMA microvalve of variant 1a. Specifications of variant 1b are similar except for the sealing behavior, which appears to be better for variant 1a due to the larger width of valve seat. Otherwise, increased adhesion of the membrane leads to a higher minimum pressure difference for the opening of the valve.

Table 5.18. Specifications of a pressure-compensated SMA microvalve of variant 1a. For the maximum flow, a theoretically estimated value is given, as the experimentally observed flow has been restricted by a fluidic resistor due to measurement reasons

Medium	gas
Mode of operation	n.o.
Size	20 x 11 x 3 mm^3
Diameter of valve chamber	4 mm
Maximum flow	7,000 standard ccm
Maximum pressure difference	450 kPa
Closing time	0.5–1.2 s
Opening time	1–2 s
Power requirement	< 450 mW

No systematic tests on the lifetime of valve variant 1 have been performed so far. However, since the actuation mechanism used is based on an R-phase transformation with maximum strain of 0.8%, no fatigue effects are expected even for larger cycle numbers above 10^6.

Concerning switching times and power consumption, the relatively large size of the beam elements are disadvantageous. Possible applications arise in fields where high energy densities are demanded but switching times play a secondary role. One example is gas chromatography, where reproducibility of switching is in the front and a switching time of 1 s is acceptable [270]. Shorter switching times and lower switching power require SMA actuators of smaller size. In this respect, the use of microactuators made of SMA foils and thin films is of particular interest, especially since required maximum flows are often below 1,000 standard ccm in microsystems applications.

The main specifications of SMA microvalves of variants 2 and 3 are listed in Table 5.19.

Table 5.19. Specifications of SMA microvalves of variant 2 and 3. The indicated size is achieved by laser-welded designs. By using anisotropic conduction strips, the size is 8.4 x 11 x 3 mm^3

	Variant 2	Variant 3
Medium	gas	water
Mode of operation	n.o.	n.o.
Size	5.5 x 5.5 x 3 mm^3	5.5 x 5.5 x 3 mm^3
Diameter of valve chamber	2 mm	2 mm
Maximum flow	470 standard ccm	13 ml/min
Maximum pressure difference	500 kPa	270 kPa
Closing time	20 ms	35 ms
Opening time	50 ms	50 ms
Power requirement	< 90 mW	< 90 mW

Pressure-compensated valves of variant 2 feature high controllable pressure differences up to 500 kPa and moderate flows up to 500 standard ccm. In order to keep microactuator and flow medium separated, valves of variant 3 are not pressure-compensated and, thus, show reduced controllable pressure differences.

The use of a metallic cooling ring makes a large impact on the dynamic valve performance. In this way, typical opening and closing times of 20 and 50 ms are achieved upon gas operation, which correspond well to simulations of heat transfer times. Operation with water gives rise to somewhat larger closing times due to enhanced heat loss during electrical heating of the SMA actuator. The relatively long cooling time of 50 ms is mainly due to the low temperatures of R-phase transformation.

First lifetime tests on valves of variant 2 up to 10^6 cycles reveal no deviation from the starting performance.

Compared to variant 1, significantly shorter switching times are obtained for comparable pressure differences. Thereby, applications with higher demands on valve dynamics become also available. One example for variant 2 may be the use as a pilot valve in pneumatics, which requires maximum frequencies on the order of 10 Hz [271]. Possible applications of variant 3 arise in the fields of chemical and biochemical analytics.

The specifications of SMA microvalves of variant 4 are sumarized in Table 5.20. The main differences between the subvariants are the maximum controllable pressure differences and flows, as well as the maximum ambient temperatures for operation. NiTi thin film actuators with R-phase transformation exhibit relatively small flows and maximum controllable pressure differences. However, the NiTi thin films used show a very ductile behavior, which facilitates hybrid integration.

Table 5.20. Specifications of SMA microvalves of variant 4

	NiTi thin film actuator	NiTiCu thin film actuator	NiTiPd thin film actuator
Medium	N_2 gas	N_2 gas	N_2 gas
Mode of operation	n.o.	n.o.	n.o.
Size	6 x 11 x 3 mm^3	6 x 11 x 3 mm^3	6 x 11 x 3 mm^3
Diameter of valve chamber	2 mm	2 mm	2 mm
Maximum flow	180 standard ccm	360 standard ccm	360 standard ccm
Maximum pressure difference	200 kPa	300 kPa	270 kPa
Power requirement	< 30 mW	< 100 mW	< 210 mW
Maximum ambient temperature	35 °C	20 °C	60 °C

The best fluidic specifications among microvalves of variant 4 are achieved for NiTiCu thin film actuators displaying a martensitic transformation. However, the martensitic transformation is only fully used for ambient temperatures of 20 °C and below. In contrast, microvalves with NiTiPd thin film actuators allow the largest temperature range of application. The given limit of 60 °C can be further extended to higher temperatures by increasing the Pd content in the thin film. In this case, sacrificial layer technology may be replaced by an integration technology based on freestanding thin films in order to avoid diffusion problems during heat treatment.

Simulations demonstrate that valve dynamics are well improved by the use of SMA microactuators with higher phase transformation temperatures. Presently, further development efforts, especially on microvalves with NiTiPd microactuators, concentrate on experimental verification of the simulation results.

The maximum strain in the SMA microactuator is limited to about 0.5% by using a spacer of moderate thickness of 50 µm. The smaller spacer thickness, as well as the smaller actuator thickness, give rise to lower maximum controllable pressure differences compared with variant 2. Further improvement of pressure and flow range is possible, since the load limits of the SMA actuators are not yet reached.

Comparing the specifications of the presented SMA microvalves with specifications of other microvalves realized so far, listed in Table 2.3 (Sect. 2.2), makes the competitiveness of the SMA microvalves clear. Variant 1 already shows promising results with respect to the maximum controllable pressure differences and flows. However, the power requirement and switching times are relatively large. Variants 2 and 3 are among the smallest microvalves realized so far. Nevertheless, the controllable pressure differences and flows are comparable to the specifications of state-of-the-art electrostatic and bimetallic microvalves. The power requirement is even lower by more than an order of magnitude compared with thermopneumatic and bimetallic valves. As expected, ony the switching times are outperformed by electrostatic and piezo valves.

5.4 Normally Closed SMA Microvalves

Normally closed valves maintain a closed condition while no power is supplied. This function may be realized by a mechanically coupled system consisting of an SMA actuator with one-way effect and a passive biasing spring. The SMA actuator is used for opening of the valve. The spring is responsible for resetting the SMA actuator after the power supply is turned off. Furthermore, the spring maintains the closed condition as long as the power supply remains disconnected. In the following, the development of valve variants will be addressed, which are controlled by an SMA thin plate actuator of 100 μm thickness. In this case, the biasing spring is fabricated from an SMA thin plate with pseudoelastic effect.

5.4.1 Valve Design

The design of valve variant 5 is based on the concept of beam elements developed for valves of variant 1. SMA beams of 100 μm thickness allow large displacements associated with moderate requirements on fabrication tolerances, which can still be met by conventional precision machining. Thus, rapid prototyping of these types of valves is possible. The procedure for designing the SMA microactuator is oriented toward the optimization criteria discussed in Sect. 4.3.

The SMA microactuator, as well as the biasing spring, are designed as circular arrangements of stress-optimized beams. Fig. 5.39 shows the lateral shape of both devices. Direct heating of the SMA microactuator by an electrical current is enabled by interconnecting the beams in series and providing two contact pads. The biasing spring is designed with circular symmetry.

Fig. 5.40 shows two types of valve variant 5. The main components are an upper and a lower housing with integrated fluidic interconnections, a ring, a plunger with a sealing, an SMA microactuator, and a pseudoelastic SMA microspring. The underlying ideas of the valve setup may be summarized as follows:

- Valve membrane and spacer of previous valve variants are replaced by a ring and a freely movable plunger. Among other functions, the ring guaranties a guided motion of the plunger. The difference in length between plunger and ring determines the pre-strain of microactuator and microspring.
- To reduce the fabrication effort, only a valve opening is provided instead of a stepped valve seat. The valve opening is closed by a sealing of silicone attached to the plunger.
- In variant 5b the valve opening is located in the lower housing, while in variant 5a it is integrated in the ring.
- One advantage of variant 5a is that the valve properties are determined only by the fabrication tolerances of the ring and plunger. Tolerances of the lower housing are unimportant in contrast to variant 5b.

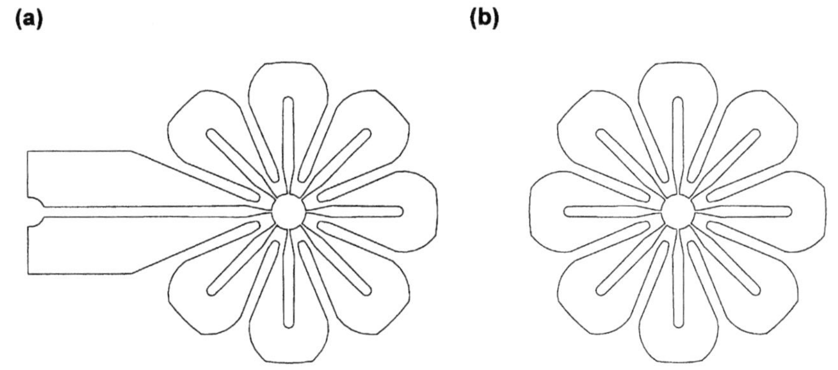

Fig. 5.39. Designs of the lateral width profile of the SMA microactuator (a) and biasing spring (b)

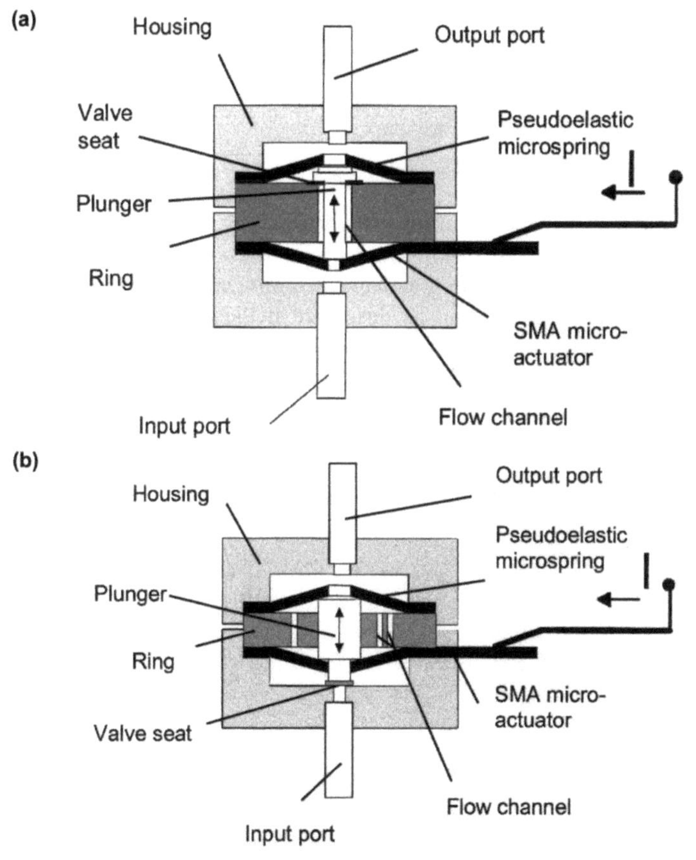

Fig. 5.40. Schematic of two types of valve variant 5

- One advantage of variant 5b is that the controllable pressure difference can be adjusted within a large range by the design of the valve opening in the lower housing. In variant 5a, the pressure acts on the whole area of the valve opening in the ring, which has to be designed with a given minimum size due to fabrication limits.
- In variant 5b the plunger has to be guided exactly to keep it centered with respect to the valve opening. Since the annular gap of the plunger strongly limits the flow, additional openings are provided in the ring.
- The openings in the ring of variant 5b allow pressure compensation between the upper and lower valve housings.

The load case in valve variant 5b is illustrated schematically in Fig. 5.41. In closed condition, the spring force F^{SE} balances the force of the SMA microactuator in R-phase condition F^R, the pressure-induced force $F_{\Delta p}$, and the closing force of the plunger F_c:

$$F^{SE} = F^R + F_{\Delta p} + F_c \,. \tag{5.13}$$

The closing force decreases for increasing pressure-induced force. The maximum controllable pressure difference is achieved when the closing force decreases to zero. For higher pressure-induced forces, the valve opens up and remains in an open condition. If the valve is opened by electrically heating the SMA microactuator, the following force balance appears:

$$F^{SE} = F^A + F_{\Delta p} \,. \tag{5.14}$$

Eqs. 5.13 and 5.14 are also valid for valve variant 5a.

The lateral shape of the beam elements is stress-optimized, as for valves of variant 1, in order to fulfill the mechanical optimization criteria (O1)–(O3). The procedure of stress optimization is described in Sect. 4.3.1. As the load on the beam can be well approximated by a point load acting at the beam end, stress optimization results in a trapezoid lateral shape.

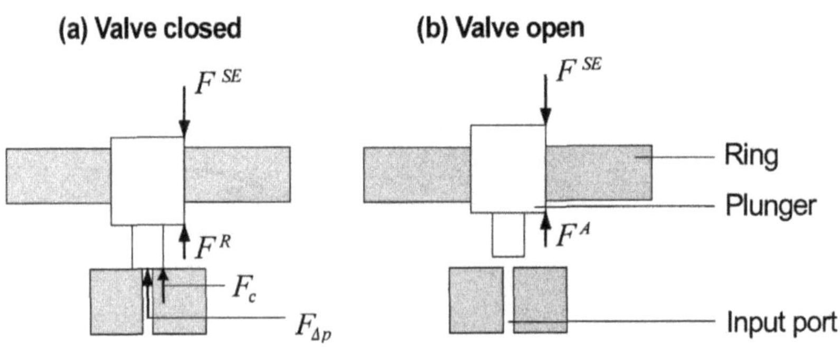

Fig. 5.41. Load case in a valve of variant 5b in closed (a) and open condition (b)

Taking into account various geometrical boundary conditions like, for instance, the diameter of the valve chamber and the gap width produced by laser cutting, a circular design of eight double beams is obtained with minimum and maximum lateral width of 100 and 600 μm, respectively. The outer diameter of the SMA microactuator and SMA microspring is set to 5.2 mm.

Fig. 5.42 illustrates schematically the progression of force-displacement characteristics of the SMA microactuator and SMA microspring. The closed and open conditions correspond to two operating points, as indicated, that are determined by the maximum stress and strain in the microdevices.

From Fig. 5.42, the following features of valve variant 5 can be deduced:

- The separation of the operating points determines the maximum stroke of the plunger.
- By increasing the force of the plunger, the controllable pressure difference increases. At the same time, the maximum fluid flow decreases, since the maximum stroke of the plunger decreases.
- Stiff springs favor large pressure differences, while soft springs favor large strokes.

The maximum stress in the SMA microactuator is determined from requirements on maximum work output and the number of actuation cycles. By making use of an R-phase transformation, the maximum strain is limited by 0.8%.

The maximum controllable pressure difference is given by the maximum closing force of the plunger F_c, which is restricted by the mechanical load limit of the microspring. The stress in the microspring is set by its design and displacement. As the lateral shape of the microspring is already determined by stress optimization, the spring constant is adjusted by its thickness. In the present case, the ratio of spring constants of microspring and microactuator in austenitic condition is designed to 1:3. The resulting thickness ratio of the microspring and microactuator is 70:90. A maximum stress in the microspring of 200 MPa results in a maximum closing force of the plunger of about 80 mN.

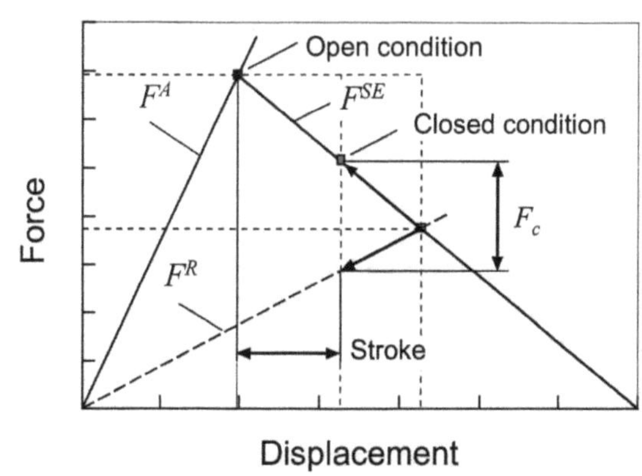

Fig. 5.42. Schematic of the force-displacement characteristics of an SMA microactuator in austenitic and R-phase conditions and an SMA microspring. The force and resulting stroke of the plunger are indicated

The main design parameters of variant 5 are summarized in Table 5.21.

Table 5.21. Design parameters of SMA valves of variant 5. For variant 5b, the length of plunger refers to the length of the guided section of the plunger responsible for the pre-straining of the microactuator and microspring

	Variant 5a	Variant 5b
Diameter of valve chamber	4 mm	4 mm
Diameter of valve opening	1 mm	0.3 mm
Diameter of sealing	1.4 mm	0.5 mm
Ring thickness	1 mm	0.4 mm
Dead volume	17.5 μl	17.5 μl
Diameter of inlet	0.6 mm	0.3 mm
Diameter of outlet	0.6 mm	0.6 mm
Length of plunger *	1.32 mm	0.56 mm
Minimum beam width	100 μm	100 μm
Maximum beam width	600 μm	600 μm
Beam length	1.75 mm	1.75 mm
Number of beams	16	16
Thickness of SMA microactuator	90 μm	90 μm
Thickness of SMA microspring	70 μm	70 μm

* In the case of variant 5b, the length of plunger only refers to the length of the guided section required for pre-straining the microactuator and microspring

5.4.2 Valve Fabrication

For rapid prototyping, the upper and lower housings, plunger, and ring are fabricated by conventional precision machining of a polymer. However, parallel fabrication of the components in batch mode by molding is easily possible. Fabrication of the sealing is performed by a casting process, as with spacers of silicone rubber. The thickness of the SMA plates is adjusted by chemical polishing. Later on, micromachining of the SMA plates is performed by laser cutting.

The valve is subsequently fabricated by hybrid integration of the components using adhesive bonding technologies. First, the SMA microactuator and SMA microspring are mounted on the lower and upper housings, respectively. For final assembly, a fit-up aid is designed, in which the lower housing, ring, plunger, and upper housing are inserted successively. Upon pressing the upper and lower housings with respect to each other, microactuator and microspring are pre-strained by the plunger. In this condition, the contact areas of the housings are bonded.

In order to obtain the desired valve performance of variant 5b, the required tolerance of the gap between valve opening and plunger should be less than 10 μm, which is difficult to fulfill by rapid prototyping. Therefore, a procedure for active alignment has been developed. In this case, one section of the lower housing containing the valve opening is designed as a movable part with a fine pitch thread.

Thus, after bonding of the upper and lower housings, the movable part can be fine-adjusted with respect to the plunger before it is fixed by an adhesive.

Electrical contacts are made by laser welding. Fluidic interconnections are realized by adhesive bonding standardized capillaries into designated flow channels. Fig. 5.43 shows a microvalve of variant 5a. Part of the upper housing has been removed to allow a better view of the valve interior. The SMA microspring can be recognized, which pushes the plunger in the center.

5.4.3 Stationary Valve Performance

Stationary characteristics of gas flow and electrical resistance are shown in Fig. 5.44 for a valve of variant 5a. The indicated pressure values are pressure differences with respect to a reference pressure of 100 kPa.

The maximum controllable pressure difference is about 200 kPa. At this pressure difference, the valve begins to open at a critical heating power of about 100 mW. The pressure difference supports the opening process. Therefore, a decrease in pressure difference gives rise to an increase in critical heating power, which determines the progress of phase transformation and, thus, the actuation force.

In open condition, the gas flow causes an extra convective cooling of the SMA microactuator. Thus, the ongoing process of valve opening is associated with an increase of power requirement causing only a gradual increase of gas flow for increasing heating power, which does not reach its maximum even at 350 mW.

Fig. 5.43. SMA microvalve of variant 5a

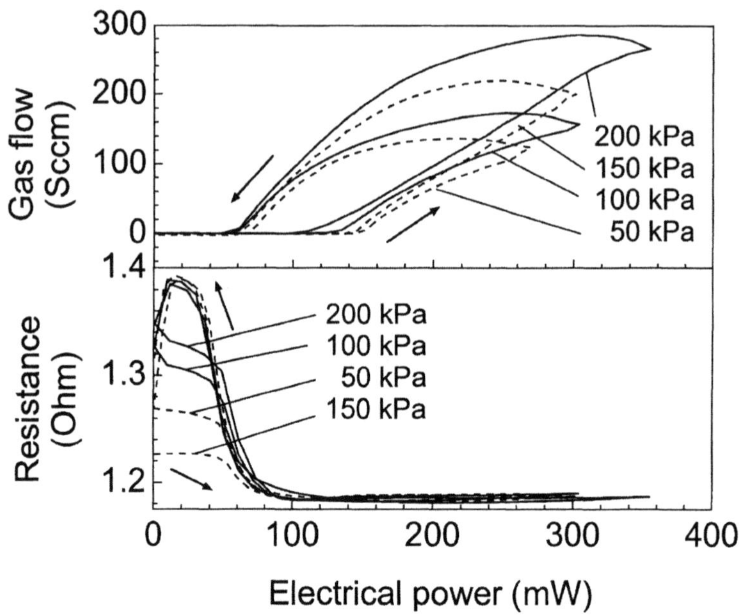

Fig. 5.44. Gas flow and electrical resistance characteristics of an SMA microvalve of variant 5a

At 350 mW, the maximum gas flow is about 280 standard ccm. By subsequently reducing the heating power, a small increase in gas flow is still observed, since extra heating of the neighboring valve components retroacts on the displacement of the SMA microactuator. The critical heating power for closing the valve upon cooling is about 60 mW. The leak rate in closed condition is below 0.1%.

Fig. 5.43 shows characteristics of gas flow and electrical resistance of a valve of variant 5b in stationary condition. The maximum controllable pressure difference is about 800 kPa. In this case, the valve begins to open above a critical heating power of about 60 mW. Also, in this case, a gradual increase in gas flow is observed by increasing the heating power, which is attributed to the simultaneous increase in cooling capacity. At a heating power of 260 mW, the maximum gas flow reaches about 460 standard ccm for a pressure difference of 800 kPa. The critical heating power for closing the valve is about 55 mW.

The resistance characteristics reflect the progression of phase transformation in the SMA microactuator. In both valve variants the behavior of opening and closing is dominated by an R-phase transformation. This is reflected in the decrease of electrical resistance for increasing heating power, as well as in the narrow hysteresis widths.

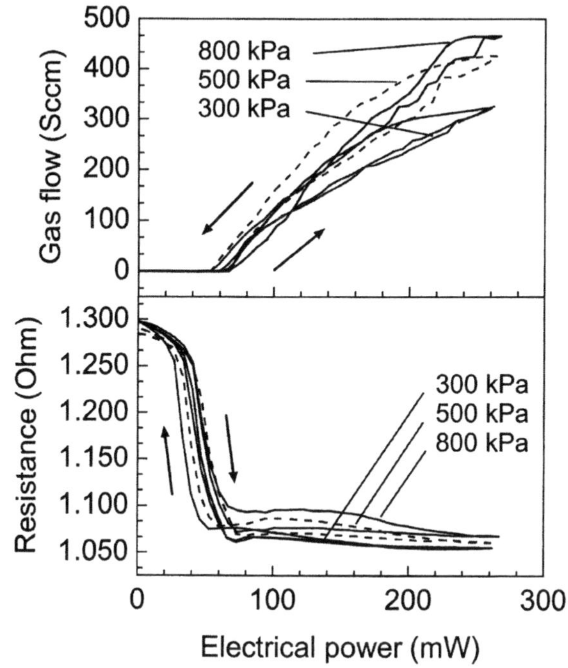

Fig. 5.45. Gas flow and electrical resistance characteristics of an SMA microvalve of variant 5b [258]

The characteristics of variant 5a shown in Fig. 5.44 already display the onset of martensitic transformation upon decreasing the electrical heating power below 10 mW, where the electrical resistance shows a steep decrease. As the M_s temperature is about 23 °C, small variations of room temperature decide on the onset of martensitic transformation. However, this has no influence on the valve function, as the valve remains closed at small heating power in any case.

The major mechanical properties of the valves are summarized in Table 5.22 for the case of maximum controllable pressure difference. The different valve openings give rise to a large difference in the maximum controllable pressure difference. Closing of valve variant 5a requires a closing force, F_c,of at least 160 mN for a pressure difference of 200 kPa. In open condition, the maximum stress in the SMA microspring is estimated to be about 400 MPa. The maximum stress in the SMA microactuator is about 250 MPa. Thus, the mechanical load limits of both microdevices are reached.

In valve variant 5b a closing force of about 60 mN is sufficient to control the pressure difference of 800 kPa. In this case, correspondingly lower maximum stress and strain values occur in the microactuator and microspring.

The corresponding fluidic properties are listed in Table 5.23. For both variants it is assumed that the flow is not fully developed and that inertial losses dominate, since the critical Reynolds number Re_t is largely exceeded for small L/D_h ratio.

Table 5.22. Mechanical properties of microvalves of variant 5. The stroke of the SMA microactuator is denoted by Δz

	Variant 5a	Variant 5b
Pressure difference	200 kPa	800 kPa
σ^A	250 MPa	125 MPa
ε^A	0.45%	0.23%
σ^R	100 MPa	55 MPa
ε^R	0.7%	0.4%
F_c	160 mN	60 mN
Δz	50 μm	35 μm

The calculated maximum flow Q is not reached by the experimentally observed gas flow Q_{exp} in the investigated range of heating power due to a simultaneous increase of cooling capacity for increasing heating power.

Table 5.23. Fluidic properties of microvalves of variant 5

	Variant 5a	Variant 5b
Re_t	74	26
Re	530	2,640
L/D_h	2	0.7
p_{in}	300 kPa	900 kPa
p_{out}	100 kPa	100 kPa
Q_{exp}	280 standard ccm	460 standard ccm
Q	1,500 standard ccm	700 standard ccm

5.4.4 Dynamic Valve Performance

The heat transfer in the microvalves is investigated as a function of time by infrared microscopy. In order to allow a direct view of the SMA microactuator, a modified setup consisting of SMA microactuator and housing is investigated, as sketched in Fig. 5.46. In this case, it is assumed that the other valve components have a minor influence on the heat transfer dynamics. Convective cooling is adjusted by the gas flow in the inlet below the SMA microactuator. For temperature measurement, the emission coefficient of infrared radiation has to be known precisely. Therefore, the surface of the SMA microactuator is covered with a thin layer of graphite, which gives rise to an emission coefficient close to 1.

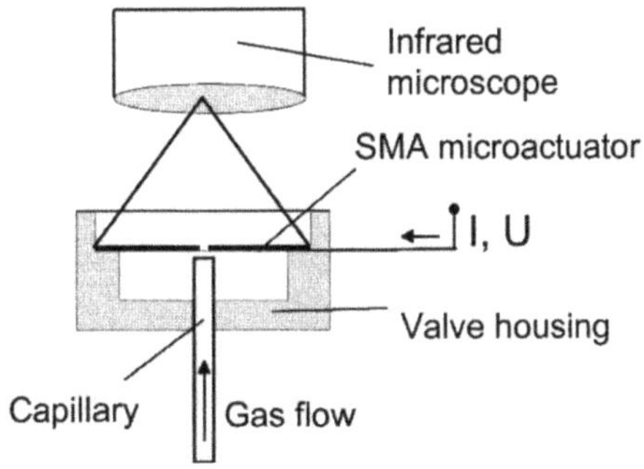

Fig. 5.46. Schematic of the setup used for infrared microscopy measurements

Fig. 5.47 shows the characteristics of maximum temperature on the surface of the SMA microactuator as a function of gas flow for various settings of electrical heating power. For a heating power of 300 mW, a maximum temperature of about 125 °C is achieved at zero gas flow. For increasing gas flow, convective cooling increases, causing a decrease in maximum temperature. For a gas flow of 600 standard ccm, it has dropped to 90 °C.

By comparing measured maximum temperatures with FEM simulations, the effective heat transfer coefficient $\overline{K}$ can be determined, which governs the convective cooling of the SMA microactuator. In the present analysis, $\overline{K}$ is used as a fit parameter, which allows a consistent fit of simulated maximum temperatures to the experimental data, as shown in Fig. 5.47 by the solid lines.

Using this procedure the characteristic of the effective heat transfer coefficient $\overline{K}$ shown in Fig. 5.48 is obtained.

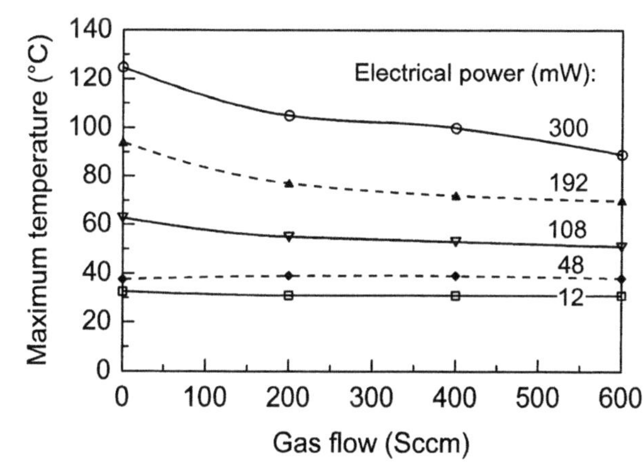

Fig. 5.47. Characteristics of maximum temperatures on the surface of the SMA microactuator of valve variant 5. The measured points are determined by infrared microscopy, the solid lines by FEM simulation using the heat transfer coefficient $\overline{K}$ as a fit parameter

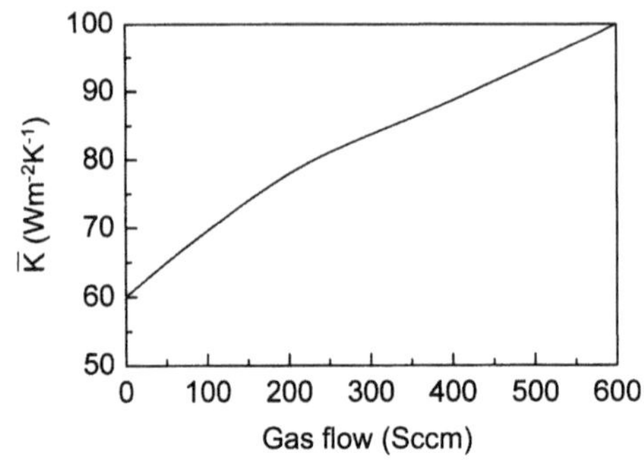

Fig. 5.48. Characteristic of effective heat transfer coefficient $\overline{K}$ as a function of gas flow

At zero gas flow, free convection is present. The corresponding effective heat transfer coefficient $\overline{K}$ is about 60 $Wm^{-2}K^{-1}$. For increasing gas flow, forced convection occurs, which is associated with an increase of $\overline{K}$. For a gas flow of 600 standard ccm, $\overline{K}$ reaches about 100 $Wm^{-2}K^{-1}$.

Fig. 5.49 shows experimental and simulated temperature profiles along the surface of the SMA microactuator for zero gas flow. The heating time is 10 s, which gives rise to a stationary temperature distribution. The electrical heating power is 300 mW. As can be seen from the numerical values at the markers, an almost homogeneous temperature profile appears along the beams. Small deviations result from the limited spatial resolution of the infrared camera of about 25 μm.

(a)

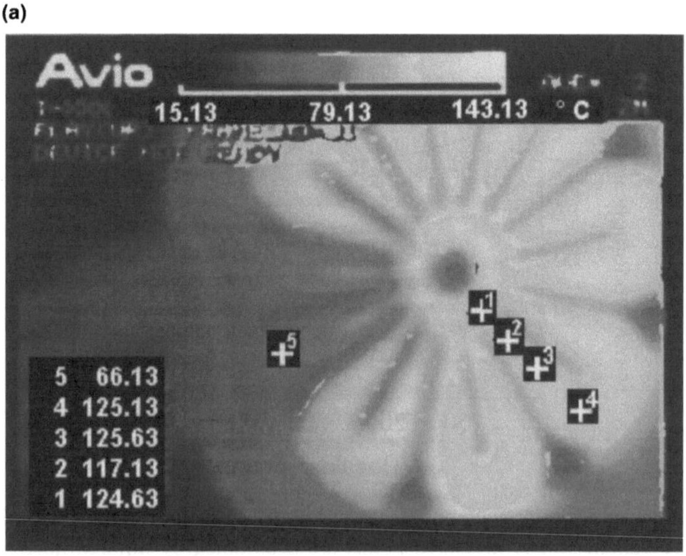

Fig. 5.49 (a) Infrared microscopy of temperature profiles on the surface of an SMA microactuator for a heating power of 300 mW at zero gas flow in stationary condition

(b)

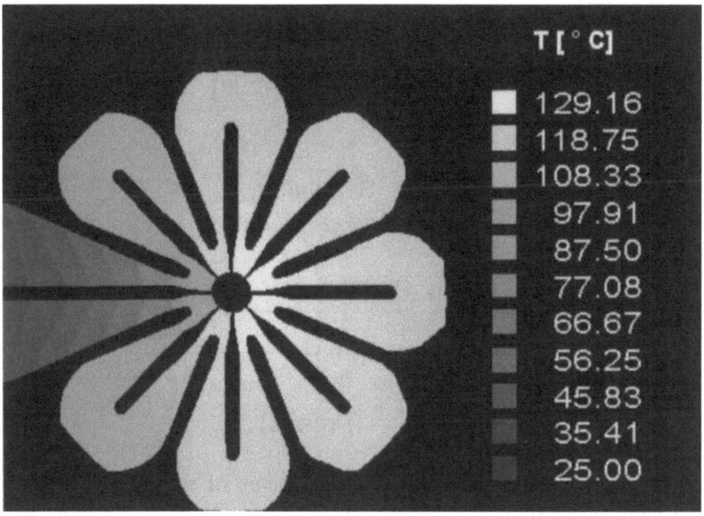

Fig. 5.49 (b) Simulation of the temperature profiles shown in Fig. 5.49(a) for a heat transfer coefficient $\overline{K}$ of 60 $Wm^{-2}K^{-1}$

The highest temperatures are found in the center of the SMA microactuator due to the small lateral width of the beam structures. Minimum temperatures occur in the beams interconnected to the electrical contact pads, which cause a relatively strong conductive cooling. Therefore, only a temperature of 66 °C is determined at marker 5.

The temperatures decrease for increasing gas flow, as shown in Fig. 5.47 for the case of maximum temperature. Simultaneously, an increasing gas flow is associated with an increasingly homogeneous temperature profile, since convective cooling equalizes temperature gradients. These features are confirmed quantitatively by the simulations.

By using pulsed heating signals, sequences of temperature profiles upon heating and cooling have been measured. A typical characteristic of maximum temperature determined from these measurements is shown in Fig. 5.50 for a gas flow of 500 standard ccm. During the heating period, the A_f temperature is reached after about 0.5 s. After about 1.2 s, all beam elements have transformed to austenite. Stationary temperature profiles occur after about 3 s. After switching the heating power off, the maximum temperature reaches the R_s temperature after about 0.5 s. Eventually, after about 1.8 s, all beam elements have returned back to R-phase condition. This time period is extended to about 5 s at zero gas flow, as no active cooling by forced convection occurs. The simulated characteristic corresponds very well to the experimentally determined behavior.

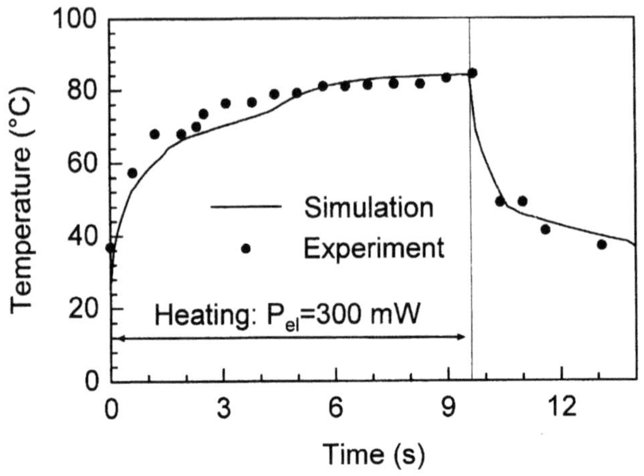

Fig. 5.50. Temporal evolution of maximum temperature of an SMA microactuator upon heating and cooling determined from infrared measurements of temperature profiles and from FEM simulations

5.4.5 Conclusions

A normally closed function is obtained in valves of variant 5 using an SMA microactuator with one-way effect and an SMA microspring with pseudoelastic effect. Both SMA microdevices consist of stress optimized beams. In the present design, a ring element is introduced in between upper and lower housings, while the membrane used for the previous valve variants is omitted. Two subvariants are presented, where the valve opening is integrated either in the ring (variant 5a) or in the lower housing (variant 5b).

The main advantage of variant 5a is the easier fabrication, as tolerances of the lower housing are unimportant. For rapid prototyping of valve variant 5b, the achievable tolerances require an active alignment procedure during valve assembly. It is expected, however, that the use of precision molding supersedes this scheme.

The specifications achieved are listed in Table 5.24. Variant 5b shows the better fluidic specifications compared to variant 5a. It is demonstrated that R-phase transformation can be used up to large pressure differences in variant 5b. So far, maximum pressure differences of 800 kPa have been realized. Even in this case, a relatively low load on microactuator and microspring is present, so that still higher maximum controllable pressure differences are feasible. Thus, variant 5b outperforms other types of SMA microvalves, fabricated, e.g., by silicon technology, as discussed in Sect. 4.5.1.

Table 5.24. Specifications of SMA microvalves of variant 5

	Variant 5a	Variant 5b
Medium	Gas	Gas
Operation mode	n.c.	n.c.
Size	6 x 6 x 6 mm^3	6 x 6 x 6 mm^3
Diameter of valve chamber	4 mm	4 mm
Maximum flow	280 standard ccm*	460 standard ccm*
Maximum pressure difference	200 kPa	800 kPa
Closing time	3.5 s	2 s
Opening time	1 s	1.2 s
Power requirement	< 400 mW	< 300 mW

*The indicated maximum flows are achieved in the given power range

The dynamic behavior corresponds to valves of variant 1, as expected, due to comparable dimensions of SMA microactuators fabricated from the same SMA material. Therefore, possible applications arise in fields where high energy densities are demanded but switching times on the order of 1 s are unproblematic. One example is gas chromatography [270].

A reduction in switching times is feasible by implementing SMA microactuators of foils or thin films, as has been demonstrated for normally open microvalves in this book. In this case, the requirements on tolerances are on the order of a few μm, which demands the application of precision molding in the development stage.

5.5 Outlook

This chapter describes a novel valve concept for microfluidics. High energy densities become accessible by using SMA microactuators with an optimized design. Furthermore, the developments demonstrate the large potential for miniaturization and high multifunctionality of SMA microactuators. For valve fabrication, new grounds of hybrid integration are broken, which, on the one hand, allow a flexible fabrication of valve components and, on the other hand, are economical due to the option of parallel fabrication. In this way, valves driven by a NiTiPd microactuator could be realized and operated for the first time. The prototypes presented may be utilized as competitive stand-alone devices. Otherwise, they may be integrated as modular devices in fluid systems of higher complexity whose development is rapidly progressing at present.

Important fields of applications that are currently being explored are for example:

- pneumatics
- chemical and biochemical analytics
- metering technology

Specific requirements of valve specifications have to be formulated based on details of applications in these fields, which will serve as a measure for dedicated developments. The valve concepts presented here will have to be further developed to open up such applications. Important objectives are, for instance:

- SMA microvalves with temperature range of application above 100 °C: This development is primarily concerned with the successful integration of thin films of NiTiPd or NiTiHf with sufficiently high transformation temperatures. Thereby, materials and interconnection technologies have to be adapted to higher temperatures. A positive side effect of this development is the improvement of actuation dynamics due to an increase in the difference between the temperatures of operation and transformation.
- Normally closed microvalves with microactuators fabricated from SMA foils or thin films: Due to the relatively small microactuator stroke, higher demands on fabrication and alignment tolerances on the order of a few μm have to be met.
- Proportional microvalves for precise control of flows: For this purpose, an additional sensoric component is required for measuring flow or size of the orifice. For optimization of control properties, SMA material systems are desired that exhibit displacement-temperature characteristics with narrow and flat hysteresis (i.e., with small temperature separation M_s/A_f and M_f/A_s, respectively, and large temperature separation M_s/M_f and A_s/A_f, respectively). In this context, SMA composites with adjustable hysteresis are of special interest [207–209].
- Capability for mass production: For realization of large quantities by parallel fabrication processes, suitable packaging and interconnection technologies on the wafer level have to be developed.
- Standardization: The realization of flexible solutions for microfluidic applications requires standardized microvalves that can be combined with other fluidic components (micropumps, sensors) in an arbitrary way. In this context, further development of modular packaging and interconnection technologies is of particular importance.

6 SMA Linear Actuators

People ask for the use of each new significant phenomenon, and they are right, as only through the use they can become aware of the value of a thing.

–Johann Wolfgang von Goethe

The unique properties of SMA microactuators appear to be very attractive for microrobotic applications. In the following, SMA linear actuators and related optical microswitches and microgrippers will be discussed. As for SMA microvalves, a dedicated technological approach is pursued for their realization, which is based on micromachining SMA plates or foils with optimized lateral shape and subsequent hybrid integration. This approach proves particularly advantageous in the combination of several microactuators to monolithic actuation systems, which can be fabricated from a single piece of SMA material in one process step. Recent publications on the development activities described in this chapter can be found in Refs. [182, 193, 272–278].

The basic setup, design, and functionality of linear actuators will be explained first. The specifications of different linear actuators will be compared with the specifications of other linear actuators presented in Chap. 2.

6.1 Actuation Concepts

6.1.1 Setup and Functionality

Linear actuation within the plane of substrate can be achieved by making use of the shape memory effect in folded beam elements of planar geometry. The basic setup of an SMA actuator producing linear motion is shown in Fig. 6.1. The actuator has a monolithic structure consisting of a central actuation unit of two four-folded beam elements, two electrical bond pads at the rear end, and an interconnecting bridge in between the beam elements at the front end. The interconnections between the beam elements are broadened to provide for a preferential direction of motion denoted as x-direction. The electrical bond pads act at the same time as mountings for mechanical fixation to a substrate. The interconnecting bridge serves for serial electrical connection of the beam elements, as well as for mechanical coupling of external forces.

In martensitic condition, the actuator can already be displaced in preferential direction by applying a small force. Thereby, the beam elements are deformed within the plane of substrate. By heating the actuator above the phase transformation temperatures to austenitic condition, the beam elements recover their

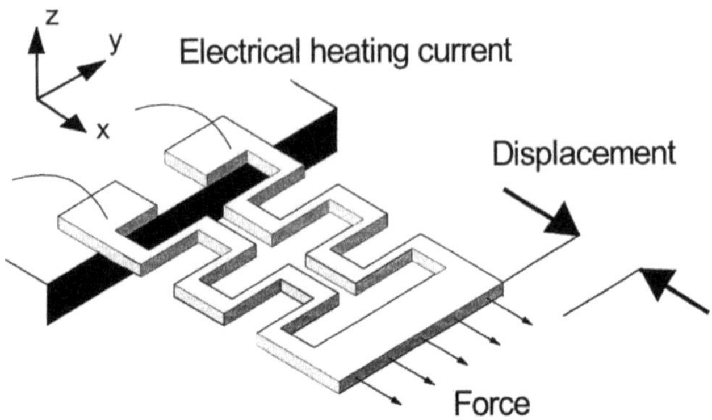

Fig. 6.1. Scheme and functionality of a SMA linear actuator

undeformed memory shape. This causes a linear motion of the interconnecting bridge in the x-direction, which can be used to generate a large work output.

Generation of an external force at the interconnecting bridge may be accomplished by a reset mechanism. Several options have been presented in Sect. 4.1. In the following, two different reset elements are considered:

(1) a passive biasing spring
(2) an antagonistic reset element

The advantage of passive biasing springs is that there is no need for electrical contacts. Furthermore, passive biasing springs provide a defined starting position in the absence of heating power. For the antagonistic reset element, a second SMA linear actuator with opposite actuation direction comes into question. Antagonistic mechanisms have the advantage that both SMA actuators may consist of the same material. Thus, the actuator can be fabricated in a single fabrication step by micromachining an SMA thin plate.

As for the SMA actuator, the structure of the biasing spring or antagonist may consist of folded beam elements. Particularly, the antagonistic mechanism suggests itself to couple frontally two SMA actuators with mirror symmetry in order to achieve the same functional properties in both moving directions.

The operation principle of an SMA linear actuator with passive biasing spring is illustrated schematically in Fig. 6.2. The linear actuator is first pre-deformed by applying a force at one end in the x-direction and is subsequently fixed at both ends on a substrate. In the absence of heating power, the biasing spring provides a defined starting position of the SMA actuator. By electrical heating, the SMA actuator recovers its undeformed memory shape and, thus, deforms the biasing spring. Turning off the heating power recovers the starting position. During such an actuation cycle, the center of the linear actuator between the SMA actuator and biasing spring performs a periodic linear motion.

As a matter of principle, replacing the passive biasing spring by a second SMA actuator results in the same movements. In this case, the reset function is accomplished by the electrical heating of the antagonist.

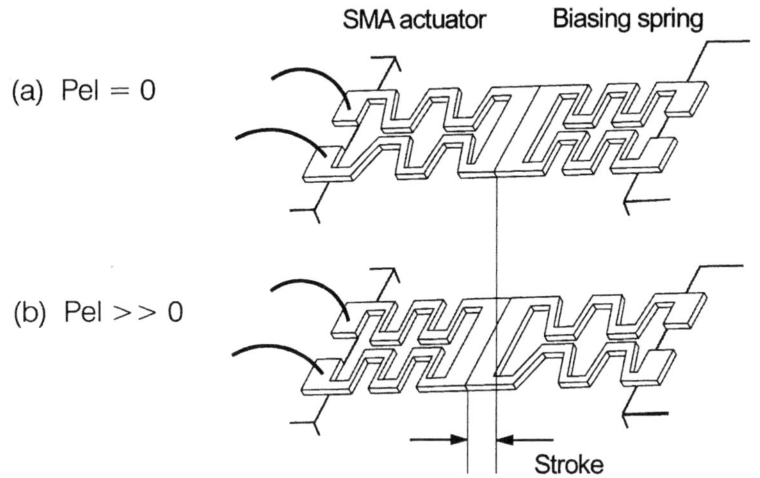

Fig. 6.2. Operation principle of an SMA linear actuator with passive biasing spring. The stroke of the center of the linear actuator is indicated by arrows

6.1.2 Design

Development activities start with available SMA materials and folded SMA beams as basic construction elements. Based on these components, SMA linear actuators are designed whose potential for applications as compact driving elements is of primary interest. The design procedure orients itself on the optimization criteria (O1)–(O6) discussed in Sect. 4.3.

The design of SMA linear actuators is largely facilitated by the use of monolithic structures. For a rough design of actuator geometry, analytic estimates based on the following assumption are made:

- The states of complete phase transformation are described by a linear model using an effective E-modulus, since reproducible strain and stress states are present in this case.
- The dynamics of linear actuation is determined by the heat-transfer times upon heating and cooling. Therefore, the actuator performance can be treated as quasi-stationary, neglecting inertia effects.

6.1.2.1 Mechanical Properties

In the simplest case, the geometry of the beam elements is designed with a constant width. However, this approach has various disadvantages, which become obvious by analyzing the stress profile upon loading. Fig. 6.3 shows a typical profile of von Mises stress in a beam element of constant width calculated by FEM simulations.

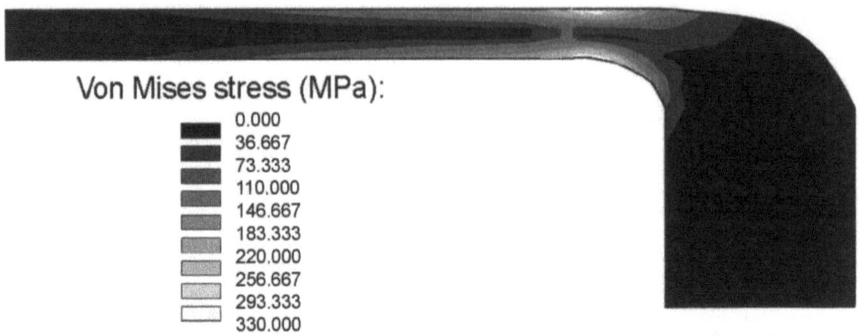

Fig. 6.3. FEM simulation of von Mises stress profile at the surface of a beam element of constant width in the region of beam support. Stress-free regions are indicated in black, regions of maximum stress in white [182]

In this case, maximum stress intensities occur only very locally at the sidewalls close to the beam supports. Consequently, the shape memory effect is only used in small volume fractions of the actuator, resulting in a relatively low work density. Furthermore, the regions of maximum stress are very sensitive with respect to overloading and fatigue.

In order to fulfill the optimization criteria (O1)–(O3), the lateral width of the beams is optimized by using the modified CAO method presented in Chap. 4. Fig. 6.4(a) shows a model of the load case in a beam element. Based on the stress profile in Fig. 6.4(b), defined modifications of geometry of certain swelling layers at the sidewalls of the beam element are calculated. These modifications are illustrated in Fig. 6.4(c) by arrows.

Fig. 6.5 displays a typical progression of simulation showing different stages of optimization. The reinforcement of beam width in regions of increased stress is clearly visible. Increasing beam width results in decreasing stress. Simultaneously, a homogeneous stress profile attunes along the sidewalls. After the maximum stress has fallen below a given stress limit σ_{limit}, the optimization procedure is terminated. For the example presented in Fig. 6.5, a stress limit of 60 MPa has been assumed.

6.1.2.2 Thermal Properties

The SMA linear actuators are operated by direct heating with an electrical current. The thermal properties resulting thereof are characterized by coupled FEM simulation. In this case, convective cooling is described by an effective heat transfer coefficient as for SMA microvalves. As the temperature range considered here is below 200 °C, the influence of heat radiation on the cooling performance is neglected.

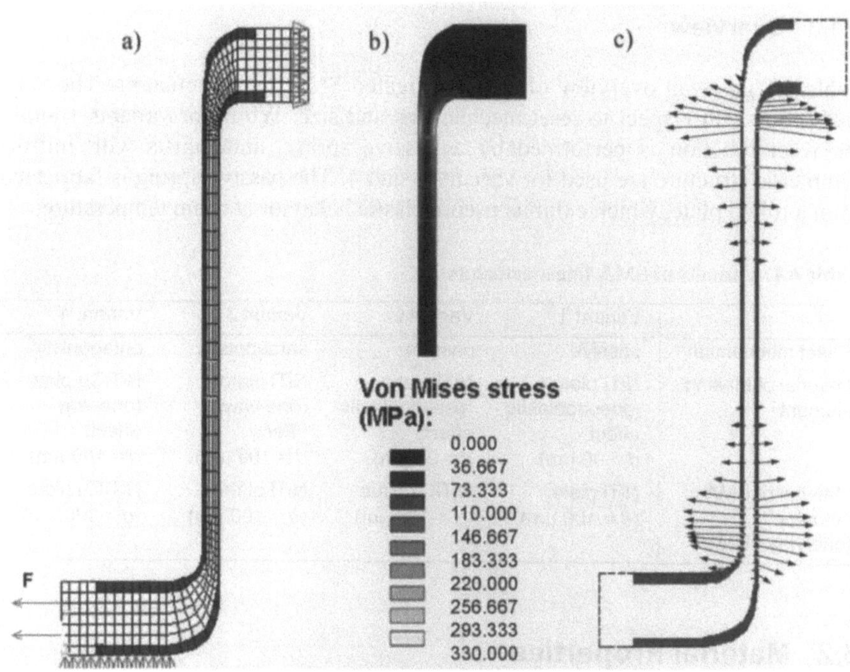

Fig. 6.4 (a) Two-dimensional FEM model for optimization of the lateral geometry of a beam element, (b) FEM simulation of stress profile shown in Fig. 6.3, (c) modifications of geometry illustrated by arrows. The swelling layers are marked in (a) and (c) in black

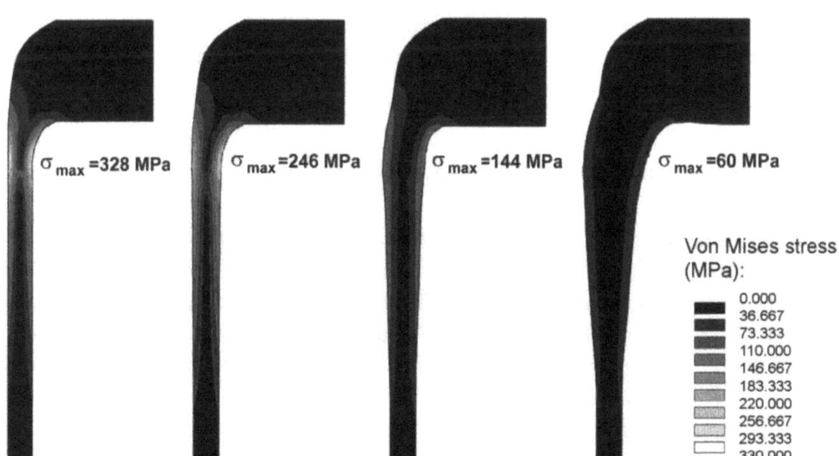

Fig. 6.5. Different stages of geometry optimization and corresponding von Mises stress profiles for the FEM model illustrated in Fig. 6.4. The maximum von Mises stress is denoted by σ_{max}

6.1.3 Overview

Table 6.1 gives an overview of the investigated SMA linear actuators. The variants differ with respect to reset mechanisms and size. While for variants 1 and 2 the reset function is performed by a passive spring, antagonists with mirror-symmetric structure are used for variants 3 and 4. The passive spring is fabricated from a rolled plate, which exhibits pseudoelastic behavior at room temperature.

Table 6.1. Variants of SMA linear actuators

	Variant 1	Variant 2	Variant 3	Variant 4
Reset mechanism	passive	passive	antagonistic	antagonistic
Material of biasing element	NiTi plate (pseudoelastic effect, *d* = 90 μm)	NiTi plate (pseudoelastic effect, *d* = 90 μm)	NiTi plate (one-way effect, *d* = 100 μm)	NiTiCu plate (one-way effect, *d* = 100 μm)
Material of SMA actuator (one-way effect)	NiTi plate (*d* = 100 μm)	NiTiCu plate (*d* = 100 μm)	NiTi plate (*d* = 100 μm)	NiTiCu plate (*d* = 100 μm)

6.2 Material Properties

For the fabrication of linear actuators, rolled thin plates of NiTi and NiTiCu are used, which can be produced in a large range of thickness. Thus, micromachined beam elements can be realized for generating displacements within the plane of substrate, whose work output is largely adjustable. The fabrication of the plates is addressed in Sect. 4.4.3. The material properties of the NiTi plates used can be found in Sect. 5.2. In the following, the main thermal and mechanical properties of NiTiCu thin plates are summarized.

6.2.1 NiTiCu Thin Plates with One-Way Effect

Linear actuators of variants 2 and 4 are fabricated from cold-rolled thin plates of NiTiCu of 100 μm thickness. The maximum cold work has been about 25% for each rolling step. The alloy composition has been determined by chemical analysis to 48.9 at.% Ni, 44.4 at.% Ti, and 5.9 at.% Cu. The oxygen content is 0.07 at.%.

First, the oxygen layer of the NiTiCu thin plate is removed. Then, the thin plate is heat treated in planar configuration at 803 K for 15 min. The material shows a martensitic transformation above room temperature. The temperatures of phase transformation determined by DSC analysis are summarized in Table 6.2. Mechanical and electrical characteristics exhibit typical hysteresis widths of about 19 K.

Table 6.2. Transformation temperatures of NiTiCu thin plates

Material	M_s (°C)	M_f (°C)	A_s (°C)	A_f (°C)
NiTiCu (Cu content 5.9 at.%) (Cold work 25%, heat treatment 803 K for 15 min)	58	46	62	76.5

By cold-rolling a preferred orientation is imprinted into the material, which leads to an orientation dependence of the transformation strain. Due to this effect, different E-moduli are determined in rolling direction (rd) and corresponding transverse direction (td). The E-moduli have been investigated by beam bending experiments. An analysis of the data for austenitic condition by an elastic beam model results in E_{rd} = 75 GPa and E_{td} = 65 GPa. Both rolling directions will contribute to the displacement of the linear actuators. Therefore, a mean value of 70 GPa is assumed for simulation. Due to the nonlinear stress-strain characteristic in martensitic condition, only strain-dependent effective E-moduli can be determined. For a maximum strain in the beam element of 1%, the effective E-modulus in martensitic condition at room temperature $\overline{E}^M$ is about 5 GPa.

6.3 SMA Linear Actuators with Passive Biasing Spring

SMA linear actuators of variants 1 and 2 consist of an SMA actuator with one-way shape memory effect and a biasing spring displaying a pseudoelastic effect. As actuator and spring generally consist of different materials, a suitable technology for joining is required. Meanwhile, technologies have been developed whereby the shape memory effect is enabled only in certain regions of an SMA component by local heat treatment [279–281]. In this case, the untreated regions may act as biasing springs, thus superseding any joining process.

6.3.1 Design and Fabrication

The structure of the SMA actuators consist of two four-folded beam elements with stress-optimized geometry. Fig. 6.6 shows an actuator of variant 2 that has been fabricated by laser cutting. The size of the actuator is about 3.5 x 2 x 0.1 mm^3. The beam width at the narrowest point is about 50 μm.

The geometry of the biasing spring is designed with the same stress-optimized width profile, since also in this case a homogeneous stress profile along the sidewalls is achieved upon loading in preferential direction. In this case, the elastic constant of the spring is adjusted by the number of folds.

(a)

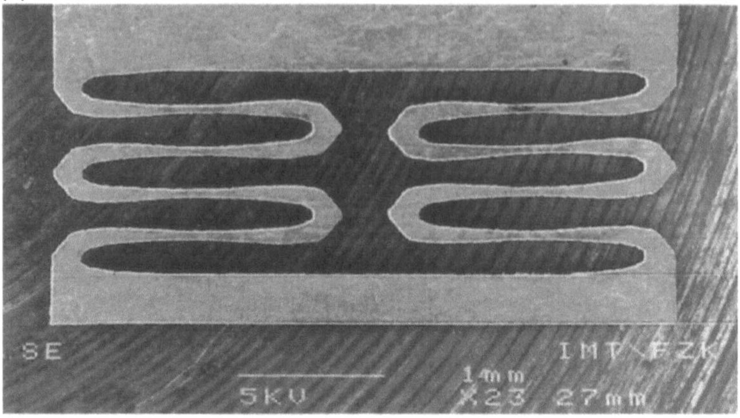

(b)

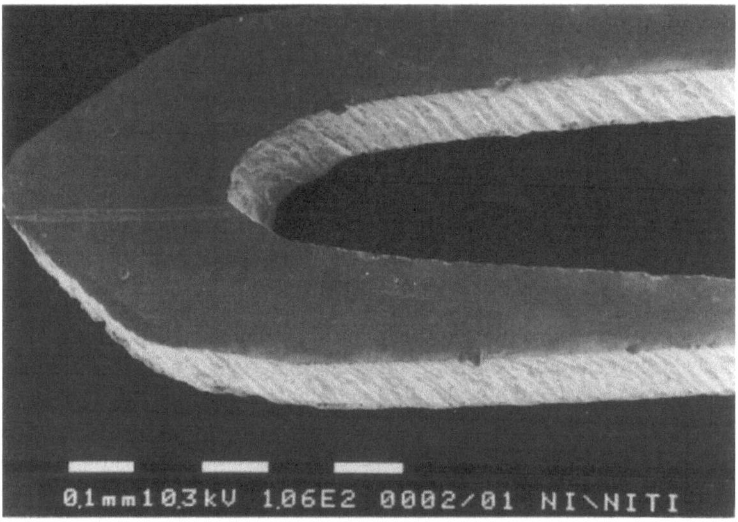

Fig. 6.6 (a) Linear actuator of NiTiCu, (b) detail of (a)

The elastic constants of the SMA actuator and biasing spring may be determined by FEM simulations of load-dependent displacements. However, the use of experimentally determined force-displacement characteristics is advantageous in order to exclude the influence of fabrication tolerances. Force-dependent displacement characteristics are measured by mounting the components in vertical orientation and adding small weights at the front end.

A linear schematic of the force-displacement characteristics of the SMA actuator and biasing spring is shown in Fig. 6.7.

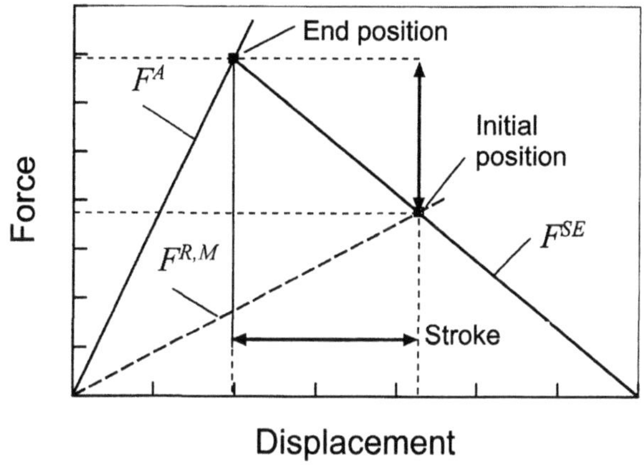

Fig. 6.7. Schematic of force-displacement characteristics of an SMA actuator and a biasing spring. F^A, F^M, and F^R denote the forces of the SMA actuator in austenitic, martensitic, and R-phase condition, respectively. F^{SE} denotes the force of the biasing spring

Based on these characteristics some features of the performance of the linear actuator can be deduced. The intersections of the characteristics determine the start and end positions of the linear actuator as indicated. These positions are determined by predeflection of the linear actuator before mounting. In the absence of current, the linear actuator is in start position; at maximum heating current, the end position is reached. From both positions, the stroke of the linear actuator, as well as the maximum change of force between SMA actuator and biasing spring can be determined.

Reasonable predeflections arise from the resulting maximum stress and strain in the SMA actuator and biasing spring. For variant 1, the maximum strain in the SMA actuator is limited to 0.8% due to R-phase transformation. For variant 2, a strain limit in martensitic condition of 1.5% is assumed, in order to avoid fatigue effects. Furthermore, the strain limit in austenitic condition has to be considered. Test measurements on beam cantilevers in austenitic condition indicate that plastic deformation starts to occur at 0.45% strain.

In addition to the load limits of the SMA actuator, the load limit of the biasing spring has to be taken into account. Conventional metallic alloys have a relatively low yield strength. Thus, biasing springs fabricated from such materials are inappropriate to achieve considerable displacements and reset forces for a given size. Therefore, an SMA material is selected that exhibits a pseudoelastic effect at room temperature. In this case, maximum strains on the order of 1.5% are uncritical.

Within the limits of a linear description, simple conditional equations can be set up for analytic estimates on predeflection of the linear actuator s_{tot}, and forces and displacements of the SMA actuator and biasing spring. The SMA actuator may be either in austenitic or R-phase (variant 1), or martensitic condition (variant 2). The displacements are denoted by s^A, s^R, and s^M, correspondingly. The predeflection of the linear actuator determines the displacements of the SMA actuator and biasing spring s^{SE}:

$$s_{tot} = s^{A} + s^{SE} \text{ (austenitic condition)} \tag{6.1}$$

$$s_{tot} = s^{R,M} + s^{SE} \text{ (R-phase, respectively martensitic condition).} \tag{6.2}$$

Since SMA actuator and biasing spring are in force equilibrium, the corresponding displacements attune according to the elastic constants FK^{i}:

$$s^{A} FK^{A} = s^{SE} FK^{SE} \text{ (austenitic condition)} \tag{6.3}$$

$$s^{R,M} FK^{R,M} = s^{SE} FK^{SE} \text{ (R-phase, respectively martensitic condition).} \tag{6.4}$$

The interconnecting bridges of the SMA actuator and biasing spring are joined to each other either by laser welding or bonding technology. In the case of laser welding, it is advantageous to perform the welding process before micromachining. As an example, Fig. 6.8 shows a linear actuator of variant 1 with a laser-welded joint. In the interconnecting bridge of the SMA actuator, small slits have been designed to thermally decouple both components during operation. In the case of adhesive bonding, both components are joined to a polyimide foil placed in between, which additionally provides for the desired thermal decoupling.

Predeflection of the linear actuator is performed by using a precision table. Then, the bond pads are fixed to an alumina substrate. Electrical contacts are made by wire bonding.

The main design parameters of variants 1 and 2 are summarized in Table 6.3.

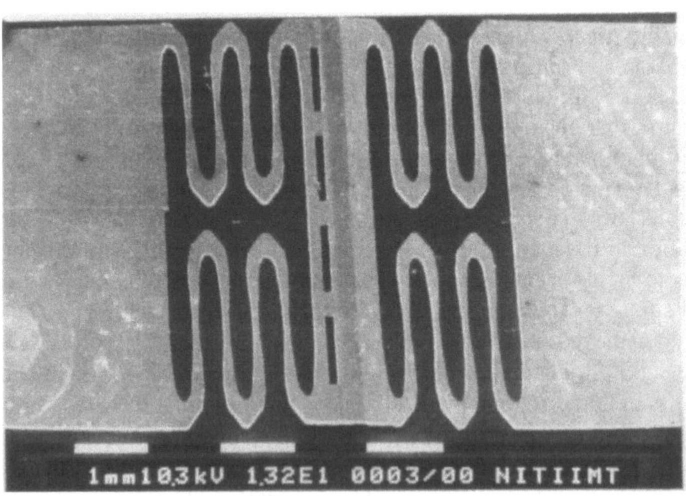

Fig. 6.8. SMA linear actuator of variant 1 with a laser-welded joint between the SMA actuator and biasing spring

Table 6.3. Design parameters of SMA linear actuators of variants 1 and 2. The effective elastic constant in R-phase condition (variant 1) corresponds to an effective E-modulus of 12 GPa. The effective elastic constant in martensitic condition (variant 2) corresponds to an effective E-modulus of 5 GPa

	Variant 1	Variant 2
Lateral size (without bond pads)	5.5 x 5.5 mm^2	3.5 x 4.8 mm^2
Material of SMA actuator	NiTi	NiTiCu
Height of SMA actuator	100 μm	100 μm
Height of biasing spring	90 μm	90 μm
Length of beam element	2.5 mm	1.5 mm
Minimum width of beam element	80 μm	50 μm
Number of beam foldings of SMA actuator	4	4
Number of beam foldings of biasing spring	4	8
Elastic constant of SMA actuator (austenite)	470 N/m	620 N/m
Effective elastic constant of SMA actuator (room temperature)	80 N/m	60 N/m
Elastic constant of biasing spring	400 N/m	200 N/m

6.3.2 Performance Characteristics

In order to allow for measurements with adjustable predeflection of the linear actuator, the bond pads of the SMA actuator and biasing spring are mounted on two separate precision tables, which can be fine-positioned with respect to each other. The performance of the linear actuators is investigated in stationary condition, as well as a function of time using pulsed electrical driving signals. Complete actuation cycles are passed through by periodic heating and cooling of the SMA actuator in the temperature regime of phase transformation. Stationary equilibrium conditions are adjusted by ramping the electrical heating power step-wise. Sufficient waiting time in each step is maintained by monitoring the change of electrical resistance as a function of time. The displacements in the center of the linear actuator are determined optically by a video microscope.

6.3.2.1 Variant 1 with NiTi Actuator

Stationary performance: Fig. 6.9 shows typical characteristics of electrical resistance and displacement of a linear actuator of variant 1 for three different predeflections. Different maximum displacements are observed corresponding to the predeflections. The range of required heating power is between 100 and 200 mW. The resistance characteristics shows a steep decrease for increasing power, which is independent from predeflection, and a narrow hysteresis. These features reflect the behavior of R-phase transformation.

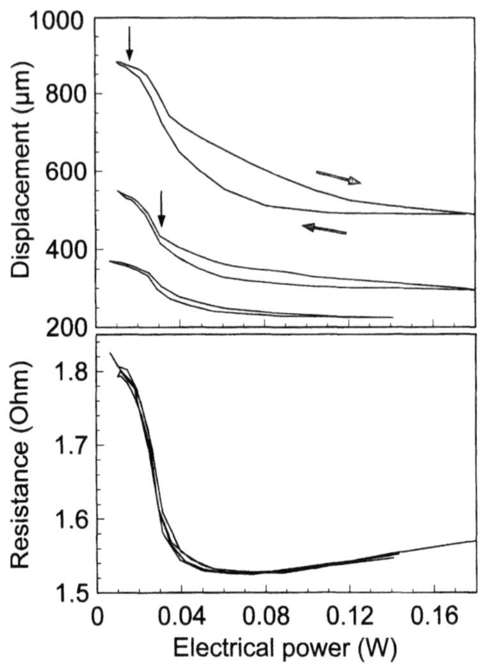

Fig. 6.9. Displacement and electrical resistance characteristics for an SMA linear actuator of variant 1. The vertical arrows indicate transitions from small to large hysteresis width

In contrast, the hysteresis width of displacement characteristics increases with increasing predeflection. At low heating power, a narrow hysteresis width is observed initially. Above critical power levels, indicated in Fig. 6.9 by vertical arrows, a considerable broadening of hysteresis sets in. This threshold behavior is observed for predeflections above 470 μm, corresponding to a maximum displacement of linear actuator of 380 μm. For increasing predeflection, the critical power level shifts from large to small values.

The threshold behavior is caused by SIM formation. For maximum displacements above about 600 μm, the hysteresis is further influenced by plastic deformation in the regions of maximum stress. Since the M_s temperature increases for increasing stress, a decrease in critical power results, above which martensitic transformation sets in. Thus, pure R-phase transformation can only occur for sufficiently low stress. FEM simulations for a predeflection of 470 μm demonstrate that the stress limit for SIM formation is at about 120 MPa. The corresponding maximum strain in R-phase condition is about 0.6%.

A heating power of 70 mW is sufficient to displace the linear actuator by about 95% of its maximum stroke. Complete actuation strokes, however, require a considerably higher heating power of 140–160 mW. This effect is caused by an inhomogeneous temperature profile of the SMA actuator particularly close to the bond pads. Experimental investigations of temperature profiles by infrared microscopy demonstrate the effect of conductive cooling by the electrical bond pads, which mainly affects the beam elements located nearby. Thus, complete phase transformation of all regions requires a significantly higher heating power.

If SIM fromation is permitted, the maximum possible predeflection of variant 1 is determined by the limit for plastic deformation in austenitic condition. For a predeflection of 470 μm, the maximum strain in austenitic condition already reaches 0.34 %. The strain limit for plastic deformation is achieved at about 620 μm.

These effects of local stress maxima are hardly observed in the resistance characteristics, since the resistance characteristics reflect the average performance of the whole component, in contrast to displacement characteristics being very sensitive for local changes of stress and strain. Thus, narrow hysteresis widths are observed in the resistance characteristics for all cases.

The mechanical properties of the linear actuators are characterized by the stress and strain values in the start and end positions. The main properties are summarized in Table 6.4 for a predeflection of 470 μm corresponding to the regime of pure R-phase transformation. The maximum von Mises stress and effective strain have been determined by FEM simulation. The maximum displacement in R-phase condition at zero heating power is 380 μm. In this case, the force between SMA actuator and biasing spring is 30 mN. In the austenitic case, the displacement decreases to 220 μm in end position. Thereby, the biasing spring is deflected accordingly, causing an increase of force to the maximum value F_{max} of 105 mN. The actuator stroke resulting from the difference of displacements in the start and end positions is 160 μm. Compared to linear actuators with constant beam width, stress-optimized linear actuators show an improvement in work density by more than a factor of 6 [182].

Table 6.4. Mechanical properties of a linear actuator of variant 1 for a predeflection of 470 μm. The terms $\overline{\sigma}$ and $\overline{\varepsilon}$ denote maximum von Mises stress and effective strain of the SMA microactuator, respectively

$\overline{\sigma}^A$	120 MPa
$\overline{\varepsilon}^A$	0.34%
$\overline{\sigma}^R$	30 MPa
$\overline{\varepsilon}^R$	0.6%
F_{max}	105 mN
F_{min}	30 mN
Actuation stroke	160 μm

Dynamic performance: The low phase transformation temperatures and the relatively large size of the SMA actuator result in cycling times on the order of several seconds. In order to achieve shorter time constants, subsequent developments aim at further miniaturization.

6.3.2.2 Variant 2 with NiTiCu Actuator

Further miniaturization of the linear actuators results in relatively small displacements and forces as long as an R-phase transformation is used. Therefore, a thin plate of NiTiCu is selected as a base material for the fabrication of SMA actuators showing a martensitic transformation above room temperature.

Stationary performance: Fig. 6.10 shows typical characteristics of electrical resistance and displacement of a linear actuator of variant 2 for two different predeflections. The corresponding maximum displacements are 310 and 610 μm.

Particularly, the displacement characteristics display a distortion at higher heating power, which is not present in corresponding characteristics obtained for a constant load. This effect is due to stress increase at the sidewalls of the beam elements of the SMA actuator, which is created by increasing displacement of the biasing spring. FEM simulations demonstrate that the maximum von Mises stress increases to about 230 MPa for decreasing displacement of the SMA actuator from 610 to 220 μm. This stress increase is associated with an increase in phase transformation temperatures by about 50 K, assuming a stress rate of 5 MPa/K.

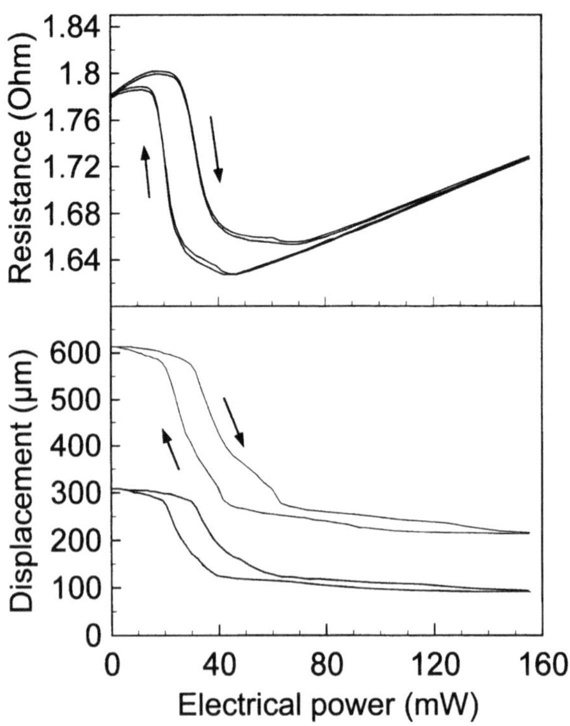

Fig. 6.10. Displacement and electrical resistance characteristics for an SMA linear actuator of variant 2

Also for variant 2, a heating power of about 70 mW is sufficient to displace the linear actuator by about 95% of its maximum stroke. However, complete actuation strokes require a much higher heating power of about 160 mW in order to compensate for the effect of conductive cooling by the electrical bond pads.

The main mechanical properties in start and end positions are summarized in Table 6.5 for a predeflection of 770 μm. In start position, the SMA actuator is in martensitic condition. In this case, a maximum displacement of 610 μm is achieved. The force between the SMA actuator and biasing spring is 33 mN. The maximum strain is determined by FEM simulations to 1.2%. In austenitic condition, the end position is reached at a minimum displacement of 220 μm. Thus, a stroke of 390 μm results in a complete actuation cycle.

The increasing displacement of the biasing spring is associated with an increase in force of up to 116 mN. The corresponding maximum von Mises stress in the SMA actuator is 230 MPa. In this case, the maximum strain in the SMA actuator is close to the load limit in austenitic condition of 0.45%. Thus, the maximum possible stroke of variant 2 is limited by the mechanical load limit in austenitic condition.

The force-displacement performance gives rise to a maximum work density of 17 mJ/g. In spite of miniaturization of lateral size by a factor of 0.625, variant 2 exhibits forces and displacements comparable to variant 1. This is enabled by using a martensitic transformation. The corresponding higher work density is, however, associated with an increased loading of components.

Table 6.5. Mechanical properties of a linear actuator of variant 2 for a predeflection of 770 μm. The terms $\overline{\sigma}$ and $\overline{\varepsilon}$ denote maximum von Mises stress and effective strain of the SMA actuator, respectively.

$\overline{\sigma}^A$	230 MPa
$\overline{\varepsilon}^A$	0.4%
$\overline{\sigma}^M$	55 MPa
$\overline{\varepsilon}^M$	1.2%
F_{max}	116 mN
F_{min}	33 mN
Actuation stroke	390 μm

Dynamic performance: The heating times required to cause a phase transformation in the SMA actuator can be determined from time-resolved electrical resistance characteristics. The procedure is illustrated in Fig. 6.11.

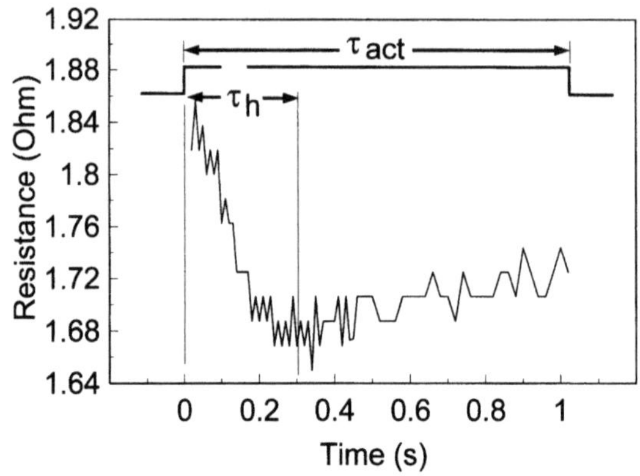

Fig. 6.11. Time-resolved electrical resistance characteristic for a linear actuator of variant 2; τ_{act} and τ_h denote actuation time and heating time required for phase transformation, respectively [277]

The heating time τ_h required for phase transformation in the SMA actuator is determined from the time interval between the onset of the heating pulse and the minimum of electrical resistance. As can be seen from the stationary electrical resistance characteristic in Fig. 6.10, the displacement of linear actuator reaches about 90–95% of the total stroke at the minimum of electrical resistance.

As shown in Fig. 6.12, the heating time τ_h decreases inversely proportional to the heating power. The solid line displays a fit function of the form $Const \cdot P_{el}^{-1}$, which is characteristic of adiabatic heating processes. In this measurement series, the electrical power is kept below 200 mW in order to avoid local overheating. For 190 mW, τ_h drops to about 250 ms. The maximum work density of 17 mJ/g and the minimum heating time result in a maximum power density of 70 mW/g.

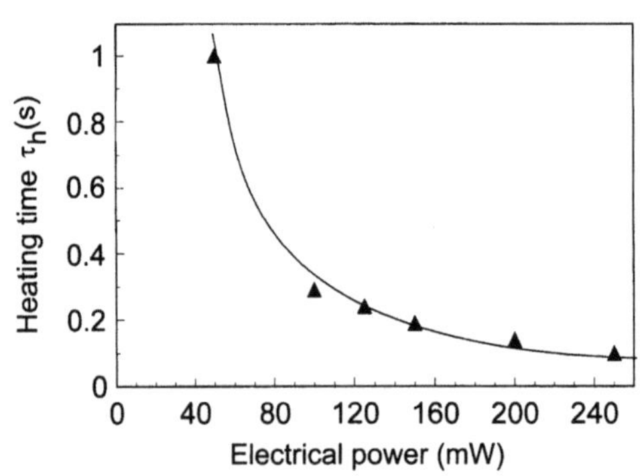

Fig. 6.12. Heating times determined by time-resolved electrical resistance measurements [277]

The cooling time that the SMA actuator requires to reach the initial temperature can hardly be determined from time-resolved electrical resistance characteristics, since the experimental accuracy becomes rather poor at low currents. Therefore, another measurement technique based on optical transmission is applied. For this purpose, a light barrier is realized by two aligned optical fibers in a face-to-face arrangement. The center part of the linear actuator is positioned with respect to the optical gap between the fiber ends in order to detect displacement-dependent transmission signals. The actual position of the linear actuator in the optical gap determines the displacement range of maximum sensitivity.

Fig. 6.13 shows a typical time-resolved transmission signal. Since the fibers used have a core diameter of 50 μm, the displacement can be resolved only partially. However, this is no problem for measurements of cooling time, as only the displacement range close to the initial position has to be resolved. Consequently, the maximum sensitivity is adjusted to this displacement range. In the remaining displacement range, the signal displays a plateau.

The cooling time τ_c is experimentally determined from the time interval between the end of the driving signal and the point in time where the linear actuator has reached its initial position. For a heating power of 190 mW and a duration of heating pulse τ_{act} of 300 ms, τ_c is about 1.5 s. Compared to variant 1 a considerable improvement in dynamics is achieved by a factor of about 3. The improved cooling time is due to the reduced size, as well as to the somewhat higher phase transformation temperatures of the SMA actuator.

The result on cooling time τ_c reflects the time interval of the complete reverse transformation. Considerably shorter switching times can be achieved if partial phase transformations are used that require only a fraction of cooling time. In this case, however, only a fraction of complete actuation stroke is exploited. Fig. 6.14 shows, for instance, periodic switching motions of a linear actuator of variant 2 for a heating signal with a 75 ms duration. During this time interval only a partial phase transformation is induced.

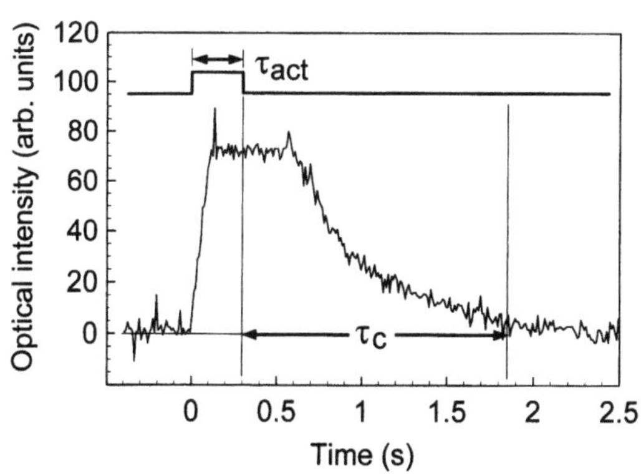

Fig. 6.13. Time-resolved optical transmission characteristic for the measurement of the cooling time τ_c of an SMA linear actuator of variant 2 [277]

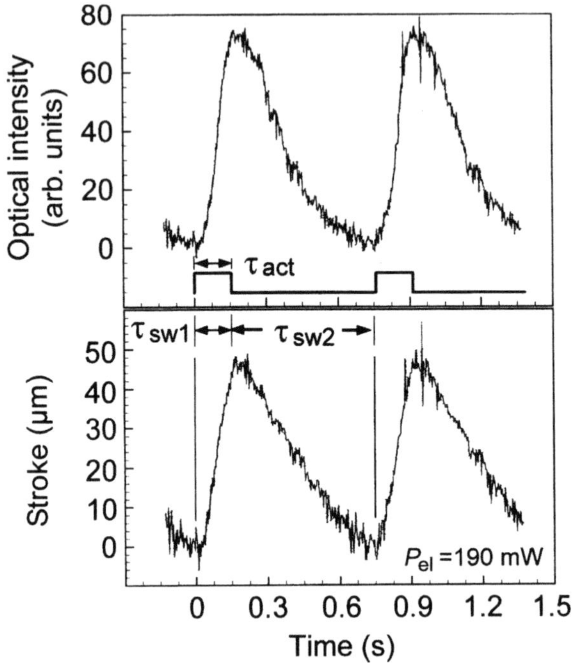

Fig. 6.14. Switching performance of an SMA linear actuator of variant 2. The power and duration of the heating pulse are 190 mW and 75 ms, respectively; τ_{sw1} denotes the switching time in actuation direction, and τ_{sw2} in reset direction

In this case, the duration of the heating pulse corresponds to the switching time in the actuation direction τ_{sw1}. Within the switching time τ_{sw1}, the linear actuator is displaced by about 50 μm. The dynamics of reset motion is determined by the cooling behavior of the SMA linear actuator. Thus, the initial position is not reached until a switching time τ_{sw2} of about 300 ms.

6.3.3 Conclusions

Linear actuators of variants 1 and 2 consist of a stress-optimized SMA actuator with a one-way shape memory effect and a stress-optimized biasing spring displaying a pseudoelastic force-displacement behavior. Variant 2 has been miniaturized with respect to variant 1 by a factor of 0.625 in order to improve the dynamic behavior. Variant 2 is furthermore made of a NiTiCu alloy instead of a binary NiTi alloy. The concept of stress optimization is used to achieve homogeneous stress profiles with an optimum stress intensity in the SMA actuator and SMA biasing spring to generate a large work density without exceeding critical values. Thus, the shape memory effect is used in an optimum way. Compared to linear actuators consisting of beam elements of constant width, an improvement in work density by more than a factor of 6 is achieved for comparable loading.

By changing to a NiTiCu actuator, the advantage of narrow hysteresis in the material characteristics of NiTi is abandoned in favor of a larger range of stress and strain. This concept succeeds in achieving comparable forces and displacements despite miniaturization. In this case, the maximum work density reaches 70 mW/g.

Fabrication of the linear actuators is reduced to only three steps: frontal interconnection of two SMA plates, micromachining, and hybrid integration. For the interconnection of the SMA actuator and biasing spring, the technologies of laser welding and adhesive bonding are pursued. Micromachining is performed by laser cutting, which is preferred for SMA plates of 100 μm thickness over etching technology due to the better aspect ratio. For hybrid integration, adhesive bonding technology is used, as for SMA microvalves. Before mounting, the SMA linear actuator has to be predeflected.

The main specifications of linear actuators of variant 2 are listed in Table 6.6.

Table 6.6. Specifications of a linear actuator of variant 2 for a maximum strain in the SMA actuator of 1.2%

Reset mechanism	Passive
Material of SMA actuator	NiTiCu
Size (without bond pads)	3.5 x 4.8 x 0.1 mm^3
Maximum reset force	116 mN
Maximum actuation stroke	390 μm
Switching time (190 mW, actuation direction)	250 ms
Switching time (reset direction)	1.5 s
Power requirement	< 200 mW

Linear actuators of variant 2 are particularly suited for the realization of large forces above 100 mN and actuation strokes up to 390 μm. For comparable size, this order of magnitude is hardly reached by other actuation principles. The switching times depend on the direction of motion due to the use of a passive biasing spring. In actuation direction, the switching time is determined by the heating time required to raise the temperature of the SMA actuator above its phase transformation temperatures. For a heating power of 190 mW, the switching time is about 250 ms. In contrast, the switching time in reset direction is determined by the cooling time, which is considerably longer, on the order of 1.5 s. Consequently, the frequency limit to operate the linear actuator with maximum stroke is mainly determined by the time constant of the cooling time. The actuation stroke decreases for driving frequencies above the frequency limit, as well as for pulse widths smaller than the heating time, since only partial phase transformations are induced.

The assets and drawbacks of SMA actuation become clear by comparing the specifications in Table 6.6 with the specifications of linear actuators based on other principles listed in Table 2.4 (Sect. 2.3). For comparable size, the work output of alternative linear actuators is outperformed by about two orders of magnitude. In contrast, the switching times of the SMA linear actuators presented are considerably longer than of comparable linear actuators. The power requirements are in the same range as that of linear actuators based on the reluctance principle.

Further improvements in switching time in the reset direction are possible by replacing the passive biasing spring by a second SMA actuator of mirror symmetry with opposite actuation direction. The resulting antagonistic mechanism should lead to a symmetric switching performance determined by the power-dependent heating time in both directions. Therefore, further developments of linear actuators address the implementation of an antagonistic mechanism.

6.4 SMA Linear Actuators with Antagonism

Both SMA actuators can be fabricated from the same material. This enables a completely monolithic structure that can be micromachined in one fabrication step. In order to compare different linear actuators with either an antagonistic mechanism or a passive biasing spring, identical SMA actuators are used. In particular, the geometries of SMA actuators of variants 3 and 4 correspond to the geometries of SMA actuators of variants 1 and 2, respectively.

6.4.1 Design and Fabrication

Fig. 6.15 shows a linear schematic of force-displacement characteristics for an antagonistic mechanism. The mirror symmetry of the SMA actuators results in symmetric characteristics. Opposite moving directions are reflected in opposite slopes. Due to predeflection of the SMA actuators, the corresponding memory shapes are reached at different operating points given by the points of zero force.

In the absence of driving current in martensitic condition, no preferred start position exists. Alternate actuation of the antagonists results in a movement of the linear actuator between the end positions 1 and 2, which are determined by the degree of predeflection. From these positions, the actuation stroke and the range of force between the antagonists can be determined. The detailed course of force-displacement characteristic in between the end positions depends on the temporal change of temperature profiles in the antagonists. If the antagonists are driven alternately keeping the complete phase transformation of the respective antagonist, the force-displacement characteristic follows the line in R-phase condition via the equlibrium point EQ^R for variant 3 and in martensitic condition via EQ^M for variant 4, respectively.

The predeflection of variants 3 and 4 is performed according to the same criteria as for variants 1 and 2. Due to R-phase transformation, the maximum strain in SMA actuators of variant 3 is limited to 0.8%. Since variant 4 makes use of a martensitic transformation, a strain limit in martensitic condition of 1.5% is assumed.

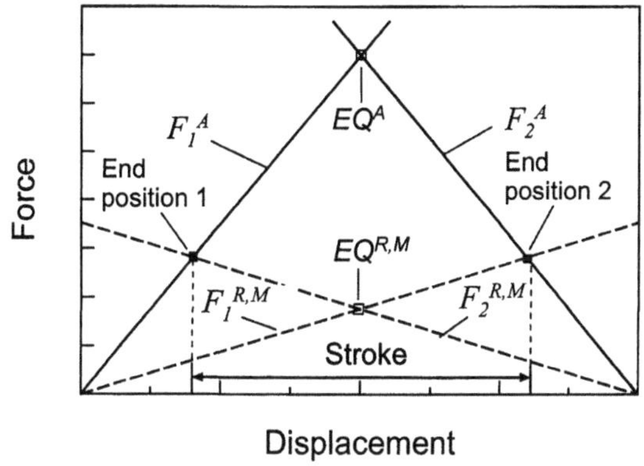

Fig. 6.15. Schematic of force-displacement characteristics for an antagonistic mechanism. The terms F^A and $F^{R,M}$ denote the forces in austenitic and rhombohedral or martensitic condition, respectively. The abbreviations EQ^A and $EQ^{R,M}$ denote the points of force equilibrium between the antagonists

In austenitic condition, the strain limit of about 0.45% has to be obeyed, above which plastic deformation occurs. Furthermore, simultaneous heating of both antagonists has to be considered, which gives rise to an additional design criterion. In this case, the antagonists are in austenitic force equilibrium EQ^A, where they should not exceed their plastic deformation limit. However, this criterion is only fulfilled for variant 3 in the case of using an R-phase transformation. In order to avoid plastic deformation in the case of large predeflection for variant 4, simultaneous heating of the antagonists has to be excluded by a suitable control mechanism.

For analytic estimates, the conditional equations 6.1–6.4 can be adapted to the case of the antagonistic mechanism. For alternating actuation, the antagonists are always in opposite phase conditions. In this case, the predeflection of the linear actuator s_{tot} equals the sum of displacements of the antagonists in austenitic and R-phase conditions for variant 3, and martensitic condition for variant 4:

$$s_{tot} = s^A + s^{R,M} . \tag{6.5}$$

The displacements attune again according to the corresponding elastic constants FK^i:

$$s^A FK^A = s^{R,M} FK^{R,M} . \tag{6.6}$$

The micromachining of both antagonists is performed by laser cutting an SMA plate in one process step. Since thermal coupling of the antagonists may have fatal consequences on the lifetime of the linear actuators, particular care is required for the design of the transition region between the antagonists.

Predeflection of the linear actuators is performed by using a precision table, as for variants 1 and 2. Then, the bond pads are fixed to an alumina substrate and electrically interconnected by wire bonding. The design parameters of variants 3 and 4 are summarized in Table 6.7.

Table 6.7. Design parameters of SMA linear actuators of variants 3 and 4. The effective elastic constant in R-phase condition (variant 1) corresponds to an effective E-modulus of 12 GPa. The effective elastic constant in martensitic condition (variant 2) corresponds to an effective E-modulus of 5 GPa

	Variant 3	Variant 4
Lateral size (without bond pads)	5.5 x 6 mm^2	3.5 x 4 mm^2
Material	NiTi	NiTiCu
Height	100 μm	100 μm
Length of beam element	2.5 mm	1.5 mm
Minimum width of beam element	80 μm	50 μm
Number of foldings	4	4
Elastic constant (austenite)	470 N/m	620 N/m
Elastic constant (room temperature)	80 N/m	60 N/m

6.4.2 Performance Characteristics

The performance of the linear actuators is investigated in stationary condition, as well as a function of time using pulsed electrical driving signals. In order to allow for measurements with adjustable predeflection of the linear actuator, the bond pads of the antagonists are mounted on two separate precision tables, which can be fine-positioned with respect to each other.

6.4.2.1 Variant 3 with NiTi Actuator

Stationary performance: In the case of the antagonistic mechanism, both SMA actuators need to be controlled successively to reach the initial position and to pass through a complete actuation cycle. Typical characteristics of displacement and electrical resistance are shown in Fig. 6.16 for a complete actuation cycle. In this case, the linear actuator has been predeflected by 1 mm.

By heating the SMA actuator I above the temperatures of R-phase transformation, the linear actuator moves to end position 1, which is reached at a minimum displacement of 190 μm. During cooling of the SMA actuator I, the linear actuator moves partially backwards due to a reduction in elastic deformation. Subsequent heating and cooling of the SMA actuator II results in a similar performance in the opposite direction of motion. Thereby, end position 2 is reached at a maximum displacement of about 810 μm. The required heating power is about 140 mW. After cooling the SMA actuator II, the linear actuator is back in the initial position. The resulting actuation stroke is about 630 μm. The maximum force between the antagonists is about 90 mN.

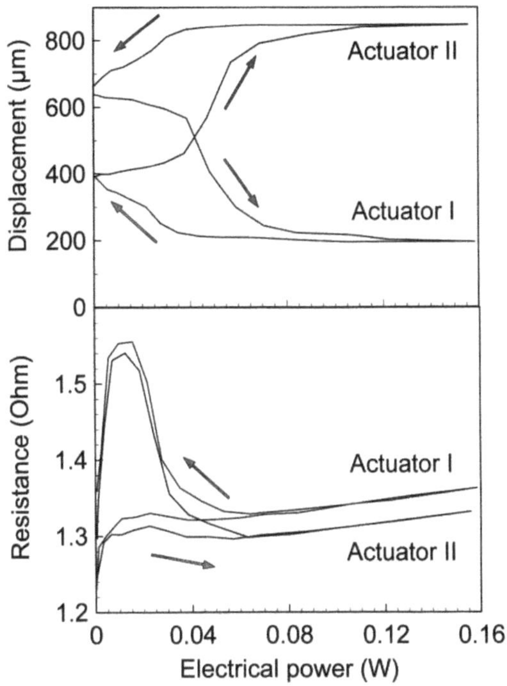

Fig. 6.16. Displacement and electrical resistance characteristics of a linear actuator of variant 3. The contributions of SMA actuators I and II are indicated [182]

The steep decrease in electrical resistance below a heating power of 10 mW indicates the onset of a thermally induced martensitic transformation upon cooling. Since the M_s temperature is at 23 °C, small changes in room temperature determine the occurrence of the transformation.

Mechanical FEM simulations in the case of room temperature indicate that the maximum strain already reaches about 1.35%. Thus, SIM formation and plastic deformation occur in addition to R-phase transformation. Both effects contribute to the broadening of hysteresis of the displacement characteristic. Furthermore, fatigue effects are present that lead, among other things, to a decrease in stroke for an increasing number of actuation cycles.

The following discussion is, therefore, restricted to the case of a maximum strain of 0.6% corresponding to a predeflection of 470 μm. In this case, a pure R-phase transformation appears, which is associated with characteristics of narrow hysteresis widths. The resulting maximum actuation stroke is about 290 μm. Thus, the corresponding actuation stroke of variant 1 of 160 μm obtained at the same strain level is largely exceeded, see Table 6.4.

This comparison demonstrates that the use of an antagonistic reset element enables a considerably larger actuation stroke compared to the use of a passive biasing spring. This result is mainly due to the small biasing force of the respective antagonist in R-phase condition, which is below 42 mN. The mechanical loading of the SMA actuators in austenitic condition is, accordingly, small. Thus, linear actuators of variant 3 are able to control considerably larger external forces than linear actuators of variant 1.

The major mechanical properties determined for the end positions are summarized in Table 6.8 for the predeflections of 470 μm and 1 mm.

Table 6.8. Mechanical properties of a linear actuator of variant 3. The terms $\overline{\sigma}$ and $\overline{\varepsilon}$ denote the maximum von Mises stress and effective strain of the SMA actuator, respectively

Predeflection	470 μm	1000 μm
Phase transformation	R phase	R phase, SIM
$\overline{\sigma}^A$	46 MPa	100 MPa
$\overline{\varepsilon}^A$	0.13%	0.3%
$\overline{\sigma}^R$	40 MPa	90 MPa
$\overline{\varepsilon}^R$	0.6%	1.35%
F_{max}	42 mN	90 mN
Actuation stroke	290 μm	630 μm

Dynamic performance: As the SMA actuators of variants 1 and 3 are basically identical, comparable heating and cooling times are obtained. The main difference in the dynamics of the linear actuators concerns the switching behavior. Due to the antagonistic mechanism, the switching times of variant 3 are determined by the heating time for both moving directions. This point will be discussed in more detail for variant 4.

6.4.2.2 Variant 4 with NiTiCu Actuator

Stationary performance: Fig. 6.17 shows characteristics of displacement and electrical resistance of an SMA linear actuator of variant 4 for two different predeflections of 300 and 600 μm. Complete actuation cycles are passed through by alternate control of the antagonists. In the present diagram, the heating power provided for SMA actuators 1 and 2 is plotted in positive and negative *x*-directions, respectively. Upon heating the SMA actuator I, the linear actuator moves to end position 1, which is reached at the minimum displacements of about 20 and 40 μm. Upon heating the SMA actuator II, end position 2 is taken corresponding to the maximum displacements of 270 and 550 μm. The hysteresis width of the characteristics reflects the specific transformation behavior of the NiTiCu alloy used.

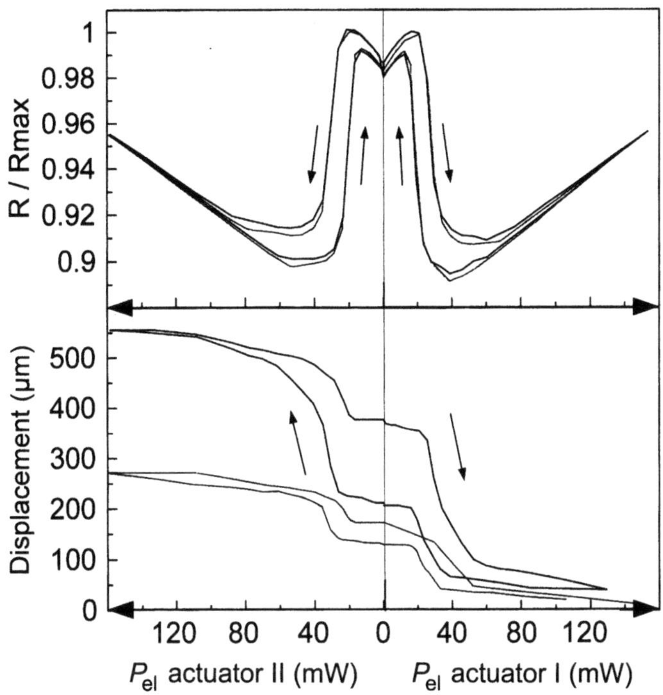

Fig. 6.17. Displacement and electrical resistance characteristics of a linear actuator of variant 4. The contributions of SMA actuators I and II are indicated [193]

A predeflection of 600 µm causes a maximum strain in the SMA actuators of 1.1%. At this strain level, the linear actuator exhibits a stroke of about 500 µm. In order to compare the results of variants 4 and 2, the mechanical properties will be discussed for a predeflection of 650 µm, which corresponds to a maximum strain of 1.2%. In this case, the actuation stroke is about 560 µm for variant 4 compared to 390 µm for variant 2, see Table 6.5. This comparison confirms the previous result for variant 3, that the use of an antagonistic mechanism is particularly advantageous for a large actuation stroke.

As for variant 3, the force between the actuating and biasing component is quite small, on the order of 30 mN. Thus, only small stress values are induced in austenitic condition in contrast to variants 1 and 2 with passive biasing springs. Therefore, the actuation stroke of variant 4 is not limited by the limit of plastic deformation in austenitic condition, as for variant 2, but by the considerably larger maximum strain in martensitic condition. Furthermore, the small force between the antagonists provides the option for control of large additional external forces.

Table 6.9 contains the main mechanical properties in the end positions for a predeflection of 650 µm, allowing a direct comparison with corresponding properties of variant 2 in Table 6.5.

Table 6.9. Mechanical properties of a linear actuator of variant 4 for a predeflection of 650 μm. The terms $\overline{\sigma}$ and $\overline{\varepsilon}$ denote maximum von Mises stress and effective strain of the SMA actuator, respectively

$\overline{\sigma}^A$	60 MPa
$\overline{\varepsilon}^A$	0.08%
$\overline{\sigma}^M$	55 MPa
$\overline{\varepsilon}^M$	1.2%
F_{max}	33 mN
Actuation stroke	560 μm

Dynamic performance: The heating and cooling times of variant 4 are comparable to corresponding times obtained for variant 2 due to the same geometry of SMA actuators. In particular, for a heating power of 190 mW, the heating time has been determined to be about 250 ms. The heating time determines the switching behavior of the linear actuator for both moving directions due to the antagonistic mechanism. Thus, a relatively fast dynamic response determined by the heating time is also achieved in the case of cyclic operation, as long as the operation frequency allows for a complete phase transformation. This situation changes, however, for frequencies higher than a critical frequency determined by the sum of the heating and cooling times. In this case, partial phase transformation occurs, which is associated with an increase in biasing force and a decrease in actuation stroke.

Experiments on switching behavior are performed by optical transmission using a light barrier. Fig. 6.18 shows, for instance, the displacement-time characteristics of a linear actuator of variant 4. In this case, the SMA actuators are driven alternately with heating pulses of 50 ms duration at a frequency of 4.2 Hz. Since the duration of heating pulses is shorter than the heating time τ_h, only a partial phase transformation is induced in the SMA actuators. The duration of heating pulses τ_{act} corresponds to the switching time in actuation direction 1 τ_{sw1} and 2 τ_{sw2}, respectively. As a result, the linear actuator performs periodic switching motions with a stroke of 20 μm compared to a stroke of 300 μm in the case of complete phase transformation. The example in Fig. 6.18 shows that switching behavior is symmetric with respect to switching direction.

The maximum stroke depends on the time interval between subsequent heating pulses τ_{rel}. A variation of τ_{rel} for a given actuation time τ_{act} leads to a characteristic change of stroke, as shown in Fig. 6.19. If the actuation time τ_{act} is longer than the heating time τ_h, the actuation stroke increases continuously as a function of τ_{rel} until a maximum value is reached, which corresponds to the actuation

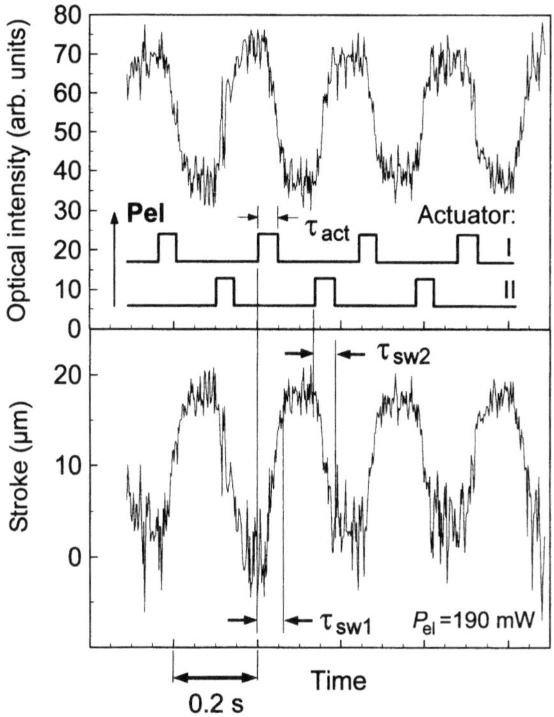

Fig. 6.18. Switching performance of a linear actuator of variant 4. The driving signals for SMA actuator I and II are indicated. The symbols τ_{sw1} and τ_{sw2} denote the switching times in actuation directions 1 and 2, respectively [277]

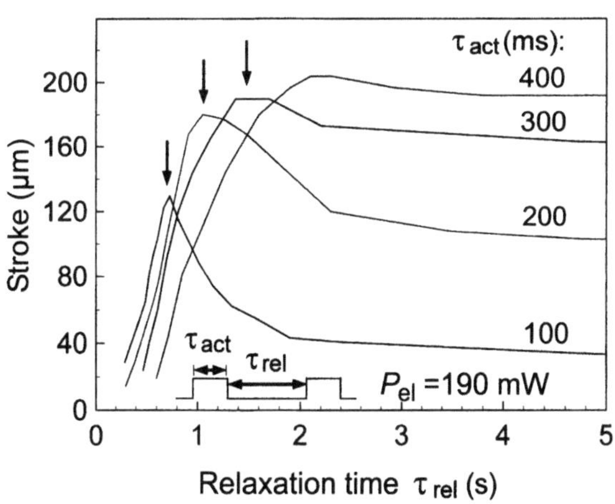

Fig. 6.19. Stroke of a linear actuator of variant 4 as a function of the time interval between subsequent heating pulses τ_{rel} for different actuation times τ_{act} [277]

stroke obtained for stationary conditions. However, for shorter actuation times $\tau_{act} << \tau_h$, a pronounced maximum of actuation stroke occurs.

This effect is due to the strong temperature-dependent change in E-modulus during phase transformation. In the range of short time intervals τ_{rel}, the effect of cooling is rather small. Therefore, the SMA actuators are essentially in austenitic condition. For increasing τ_{rel}, the average temperature of the SMA actuators decreases, and, thus, the temperature change upon heating and cooling increases, which results in an increase in actuation stroke. The effect of temperature change on actuation stroke becomes maximum once the average temperature has dropped to the regime of maximum change in the E-modulus. Time-resolved resistance measurements show that the stroke maximum coincides with the temperature regime of phase transformation, where the change of elastic material properties displays the largest gradient. This effect is only present for partial phase transformations upon heating. Once the actuation time reaches the heating time for complete phase transformation, the elastic material properties always reach their austenitic value upon heating.

6.4.3 Conclusions

Linear actuators of variants 3 and 4 consist of two stress-optimized SMA actuators that make use of the one-way effect in opposite directions with respect to each other. For operation, the SMA actuators are driven alternately with electrical current pulses. The shape of the SMA actuators of variants 3 and 4 correspond to the shapes of the SMA actuators of variants 1 and 2, respectively.

As both SMA actuators consist of the same material, fabrication is reduced to micromachining of one SMA plate and subsequent hybrid integration. The actuation center in between the SMA actuators has to be designed for sufficient thermal decoupling. For large predeflections associated with strains above 0.45%, simultaneous heating of the SMA actuators has to be excluded to prevent plastic deformation in austenitic condition.

The antagonistic mechanism used enables considerably larger actuation strokes compared to the use of a passive biasing spring. This is due to the relatively small reset force of the respective antagonist, which only slightly increases for increasing displacement, in contrast to the behavior of a passive biasing spring. The small force between the antagonists opens up reserves for the control of large additional external forces. Thus, SMA linear actuators with antagonism are suitable for switching and positioning functions that demand a large work output, but do not require a defined start position in the absence of heating power.

For variant 4, the maximum stroke is limited by the maximum strain in martensitic condition. By assuming a maximum strain in martensitic condition of 1.5%, a maximum actuation stroke of 720 µm is possible. The strokes of linear actuators with antagonistic mechanism and biasing spring (variants 4 and 2) can be compared for a maximum strain in martensitic condition of 1.2%, where variant 2 reaches its plastic deformation limit in austenitic condition. In this case, the antagonistic mechanism gives rise to an actuation stroke of 560 μm compared to 390 μm for the biasing spring. For variant 3, the maximum actuation stroke is 290 μm, which is determined by the strain limit of R-phase transformation. Apart from relatively small actuation strokes, the use of R-phase transformation is advantageous for a large number of actuation cycles above 10^6 due to negligible fatigue effects.

As a result of symmetry, the dynamic performance of variants 3 and 4 does not depend on the direction of motion. The switching times are determined by the heating times in both directions due to the antagonistic mechanism. Upon heating at a power of 190 mW, a complete phase transformation is induced within 250 ms. If no extra heat is supplied, a cooling time of 1.5 s is found. Both heating and cooling time determine the frequency limit for complete phase transformation. Partial phase transformations are induced for driving frequencies above the frequency limit, as well as for pulse widths smaller than the heating time. In this case, switching time and actuation stroke decrease. For pulsed operation, the actuation stroke can be tuned to a maximum, which occurs in the temperature regime of the largest gradient of the elastic material properties.

The main specifications of variant 4 are summarized in Table 6.10 for a maximum strain of 1.2%. Compared to variant 2 with a passive biasing spring, improvements with respect to size, actuation stroke, and switching time in the reset direction are obtained, see Table 6.6. The specifications of other linear actuators listed in Table 2.4 (Sect. 2.3) are outperformed with respect to work output by more than two orders of magnitude for the given size. However, the switching times of the SMA linear actuators are by about two orders of magnitude longer.

Table 6.10. Specifications of a linear actuator of variant 4 for a maximum strain of 1.2%

Reset mechanism	Antagonistic
Material of SMA actuator	NiTiCu
Size (without bond pads)	3.5 x 4 x 0.1 mm^3
Maximum reset force	33 mN
Maximum stroke	560 μm
Switching time in actuation and reset direction (190 mW)	250 ms
Power requirement	< 200 mW

Possible applications arise particularly in the areas of microoptics and robotics, where high demands with respect to size, displacement, and force exist. These areas will be addressed in the following.

6.5 Optical Microswitch

The strong growth of telecommunication demands the development of new technologies and devices. Multimedia applications in wide area networks require data rates of several 10 Gbit/s, which cannot be handled with conventional copper-based technology. Currently, the development trend is heading toward optical networks. For their realization, completely new devices, such as switches, routers, input and output devices and corresponding interface components, have to be to developed. One example are optical switches, which are required in optical networks, among other things, for interrupting data lines or bypassing devices in case of malfunction or testing. For these applications, low-priced devices with relatively low switching speed are of interest. In the following, two concepts of an optical microswitch based on SMA linear actuators will be presented.

6.5.1 Optical Microswitch with Movable Functional Element

One possibility for realization of an optical microswitch is based on the use of a movable optical element, which is inserted in the beam path in between two optical fibers [104, 282, 283]. In the simplest case, a beam block may be used for the optical element. For bypass applications, e.g., a micromirror is conceivable, which directs the optical beam in another direction and, thus, maintains data transfer.

Fig. 6.20 shows, for instance, an optical microswitch with an integrated beam block. On both sides of the movable beam block, two optical fibers are positioned with the optical axis aligned perpendicular to its moving direction. Depending on the position of the beam block, the optical beam path is either interrupted, partially attenuated, or deblocked.

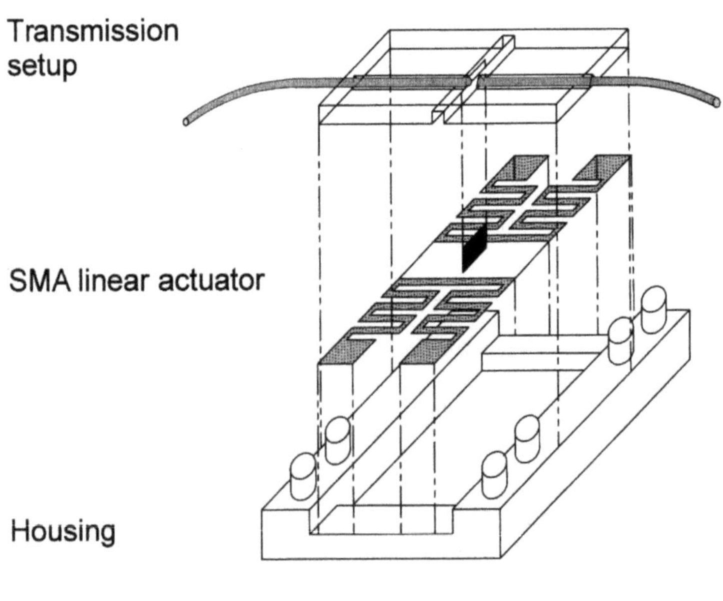

Fig. 6.20. Scheme of an optical microswitch with integrated beam block [277]

In order to keep insertion losses low, a small gap in between the fibers has to be designed. Furthermore, the end faces of the fibers have to be aligned with respect to each other without mismatch. These requirements can be met using a fiber holder with integrated fiber grooves. The fiber grooves provide a collinear guidance, as well as an alignment of the fibers in the lateral and vertical directions.

6.5.1.1 Requirements

In the case of multimode fibers with a core diameter of 50 µm, a relatively small actuation stroke below 70 µm is sufficient to interrupt or deblock the beam path. As no external forces are present, only the internal reset force has to be overcome. These requirements are fulfilled by all variants of SMA linear actuators. If value is set on compact size and frequencies below 0.5 Hz, variant 4 particularly qualifies. In the following, an optical switch with a movable beam block will be presented, which is driven by variant 4 with an antagonistic reset mechanism. In this case, a relatively small predeflection of about 100 µm will suffice to achieve the desired actuation stroke.

6.5.1.2 Fabrication

The SMA linear actuator may be fabricated monolithically, as described in Sect. 6.4.1, whereby the beam block is integrated into the actuator center in between the antagonists after micromachining. Alternatively, the optical microswitch may be assembled by joining both SMA actuators to a polyimide foil placed in between them and integrating the beam block to the polyimide foil. The second process can be easily extended to batch mode by using linear arrays of SMA actuators, which are joined in parallel to a strip of polyimide foil and separated later on.

After predeflection, the linear actuators are mounted onto the housing. A fiber holder equipped with aligned and fixed optical fibers is positioned with respect to the beam block and then mounted to the housing. The gap between the fibers is adjusted between 200 and 300 µm. Electrical interconnections to the bond pads of linear actuators are realized by wire bonding.

Fig. 6.21 shows a prototype of an optical microswitch with a movable beam block integrated into a polyimide foil. For a better view, the fiber holder has been removed. The outer dimensions of the housing are 12 x 18 x 4 mm^3.

6.5.1.3 Performance Characteristics

Displacement-transmission characteristics of the optical switch are determined by coupling the light of a LED into the lauching fiber and sensing the transmission signal in the detecting fiber by a photo sensor. The displacement of the beam block is determined optically by a microscope. Fig. 6.22 shows a typical measurement result.

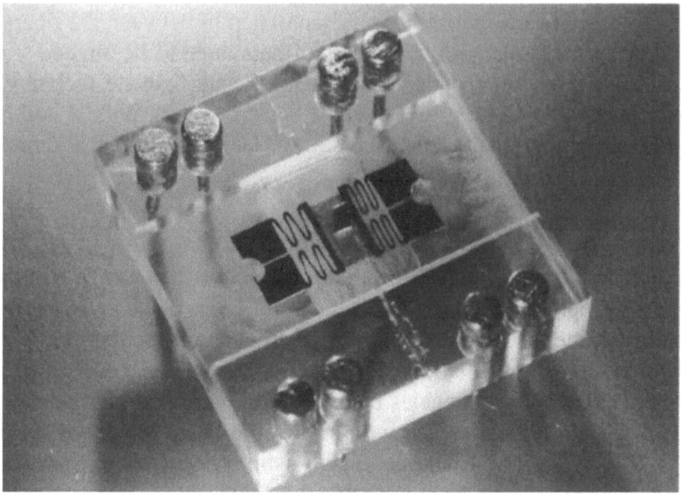

Fig. 6.21. Optical microswitch with movable beam block without fiber holder

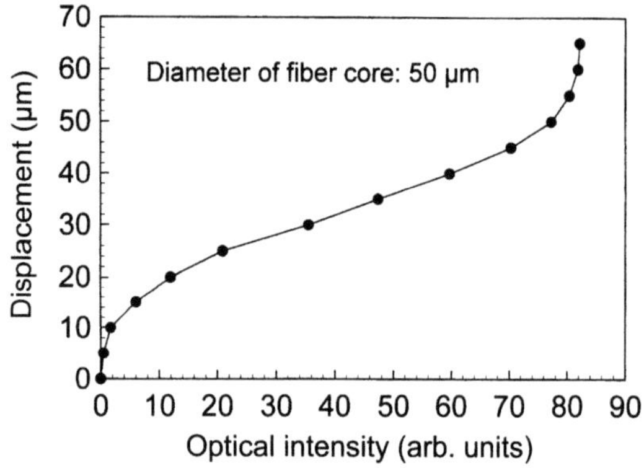

Fig. 6.22. Calibration of optical transmission signal with respect to the displacement of the beam block

An S-like shape is obtained that is determined by the coupling behavior of light into the fiber core. Since the light coupled into the cladding layer is not guided by the fiber, it does not reach the detector. In the central range of displacement between about 20 and 50 µm, the characteristic is almost linear. In this range, a maximum spatial resolution of 0.4 µm is achieved for a signal change of 1 mV.

As the gap between the fibers is relatively large, the coupling losses without a beam block are about 3 dB.

The switching behavior follows from the dynamic performance of the SMA linear actuator discussed in Sect. 6.3. Due to the antagonistic mechanism used, the switching time is determined by the heating time of the SMA actuators in both directions. The maximum frequency can be set to 1 Hz by adjusting the predeflection and making use of a partial phase transformation. By adjusting the predeflection, in particular the switching time for passing through the required displacement range of 70 μm can be reduced to at least 100 ms.

6.5.2 Optical Microswitch with Movable Optical Fiber

Another possibility for the realization of an optical microswitch is active movement of optical fibers with respect to each other, see, e.g., [284, 285]. In this case, short optical path lengths in between the optical fibers are possible, giving rise to small coupling losses. However, the displacement of optical fibers requires actuators with relatively large work output.

The operation principle of a 1 x 2 switch consisting of one input and two output channels is illustrated in Fig. 6.23. In this case, a SMA linear actuator is coupled to a translation mechanism for lateral displacement of a launching fiber. The translation mechanism consists of a movable fiber stage supported by four beams, which provide a parallel guidance. The launching fiber is fixed to the movable fiber stage. By linear displacement of the stage perpendicular to the optical axis, the launching fiber is positioned with respect to two output fibers. For a small gap in between launching and output fibers and sufficient separation of the output fibers, the optical beam path can be guided in two separate directions with low loss and low crosstalk.

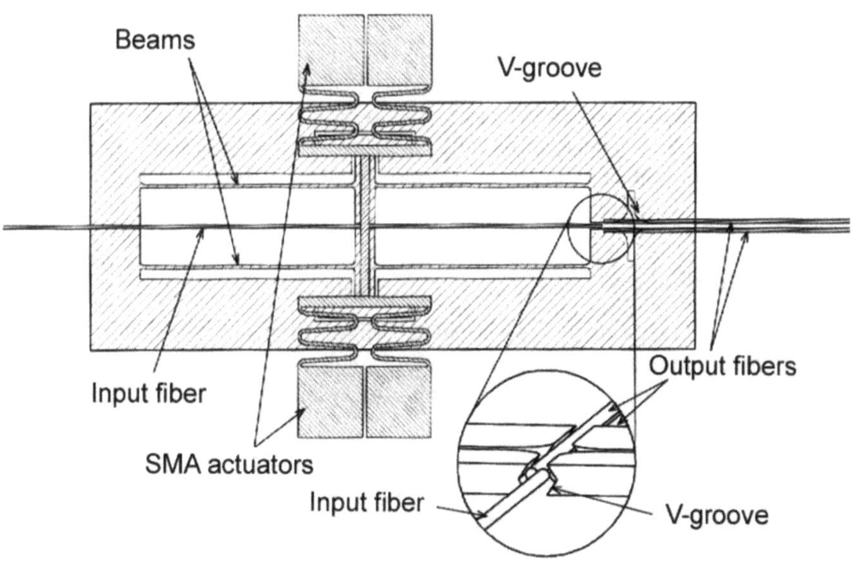

Fig. 6.23. Schematic of setup and operation principle of an optical microswitch with movable optical fiber [273]

The accuracy of this setup depends on the positioning accuracy of the movable fiber with respect to the fixed fibers. Furthermore, the separation and collinear guidance of the fibers is of importance. Stoppers are provided for the optimization of lateral positioning accuracy in the direction of actuator motion. The output fibers are fixed in vertically oriented V-grooves in order to adjust the height of the optical axis and to provide a collinear guidance. As the end of the launching fiber is not fixed and, thus, is freely bent to some extent, alignment with respect to the output fibers is performed by pressing the launching fiber to the V-groove.

6.5.2.1 Requirements

For applications in telecommunication, the design of an optical switch with maximum height of 2 mm appears particularly interesting. Therefore, the actuator should be as flat as possible. The lateral size should be in the range of a few millimeters. In order to avoid optical crosstalk, the separation of output fibers is limited to a minimum of about 200 μm. A sufficient stiffness of the movable fiber stage is ensured by designing four supporting beams with a spring constant of at least 150 N/m. Based on these guidelines, the following requirements of the SMA linear actuator follow [286]:

- Size: < 10 x 10 x 0.5 mm^2
- Force: F > 15 mN in both directions
- Displacement: s > 100 μm in both directions

No severe boundary conditions exist with respect to switching time. However, a maximum switching time on the order of 1 s is preferable. As the optical switch should take only two positions, the hysteresis of displacement characteristics plays no role.

The mentioned requirements are fulfilled by variants 2 and 4. Variant 1 exhibits too small a stroke, see Table 6.4. Furthermore, the switching times of variants 1 and 3 are longer than 1 s. In principle, both a passive and an active reset mechanism can be used. However, since equal switching positions should be realized, an antagonistic mechanism with direction-independent switching behavior appears most suitable. Therefore, variant 4 is selected.

6.5.2.2 Fabrication

The passive elements of the optical switch (fixed fiber grooves, beam elements, and movable fiber stage) are realized in Si technology. The SMA actuators are micromachined separately and integrated into the system later on. Fig. 6.24 shows a sketch of the fabrication sequence.

(1) First, a 300-μm thick Si wafer is etched through by reactive ion etching for fabrication of a through hole for the output fibers, as well as a guide hole for the launching fiber. In addition, a fixing hole for the launching fiber is etched, 213 μm deep.

(2) The sidewalls of the through hole are wet-etched anisotropically to fabricate vertical V-grooves by using a TMAH (tetramethyl ammonium hydroxide) solution.

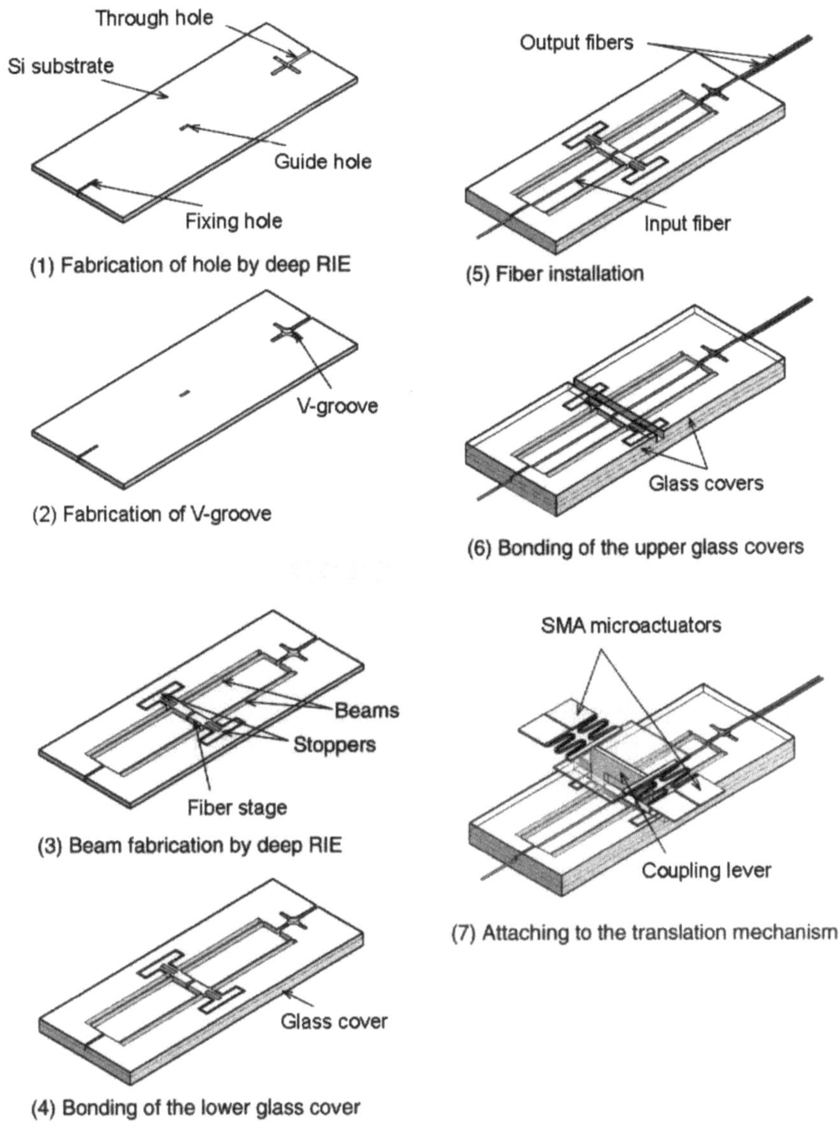

Fig. 6.24. Fabrication sequence of the optical microswitch with movable optical fiber [273]

(3) Then, the movable fiber stage, supporting beams, and stoppers are micromachined by reactive ion etching.
(4) A glass plate is bonded to the substrate from below.
(5) The optical fibers are inserted in the V-grooves and fixed by adhesive bonding.

(6) Subsequently, the upper side of the translation mechanism is also covered by a glass plate.

(7) SMA actuators are micromachined by laser cutting a 100-μm thick NiTi plate and finally, after predeflection, they are attached to the translation mechanism.

Fig. 6.25 shows an SEM micrograph of the translation mechanism. The overall size is 7 x 18 x 0.3 mm^3. The detailed view in Fig. 6.25(b) shows one of the Si beams supporting the movable fiber stage. The minimum width of the beams is 85 μm. The spring constant of the beams is determined to be 156 N/m.

The completely assembled optical switch is shown in Fig. 6.26. The lateral size is about 13 x 18 mm^2. The thickness of the device is 1.3 mm, which is considerably below the aspired size of 2 mm.

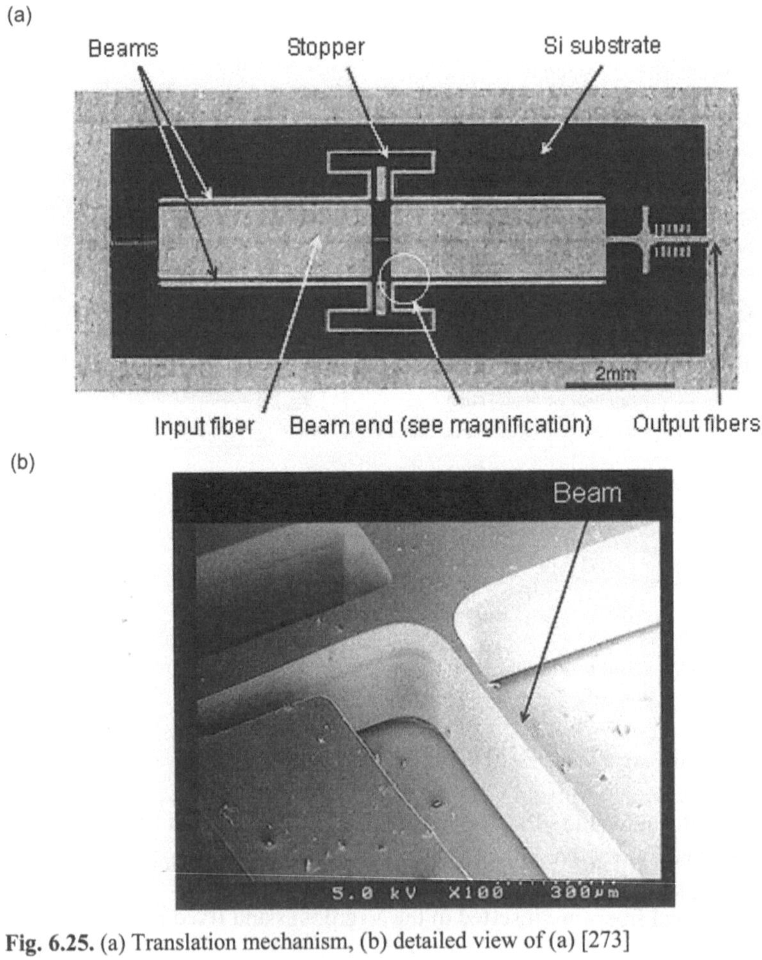

Fig. 6.25. (a) Translation mechanism, (b) detailed view of (a) [273]

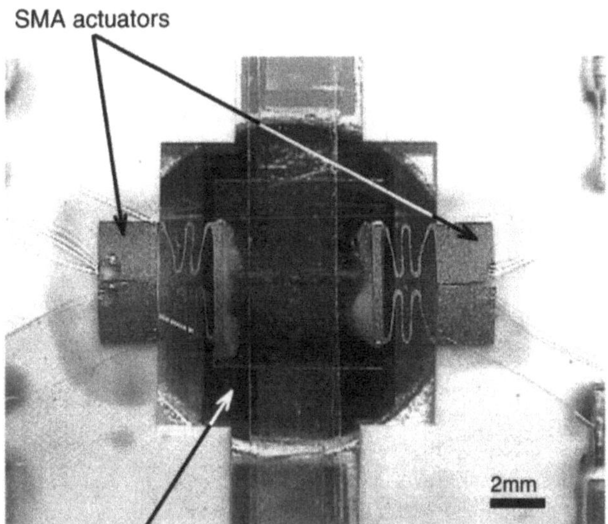

Fig. 6.26. Prototype of an optical microswitch with a movable optical fiber [273]

6.5.2.3 Performance Characteristics

The optical performance of the microswitch is mainly characterized by transmission losses. The small gap between the launching and output fibers of about 100 μm gives rise to a coupling loss of only 1.2 dB without matching of refraction index. Separation of neighboring output fibers by 200 μm is sufficient to suppress optical crosstalk by more than 60 dB.

The switching behavior is investigated by using a HeNe laser as a light source. The transmitted signal of the output fibers is measured as a function of time. The SMA linear actuators are driven alternatingly using pulsed current signals with a maximum current of 200 mA.

Fig. 6.27 shows typical time-resolved transmission characteristics. The switching time results from the time interval between the onset of current signal I_i and the point in time where maximum transmission I_l is reached. For alternating operation, this time is limited by the cooling time of the antagonist. Therefore, in the present case, the maximum possible switching time of 1.5 s is observed. For voltage-controlled operation, the time variation of current signal results from the temporal change of electrical resistance upon heating. After about 250 ms, a stationary characteristic of current is present. This time period corresponds to the heating time of the SMA actuator. The small increase at the edge of the transmitted light signal is supposed to be caused by a little misalignment of the output fibers.

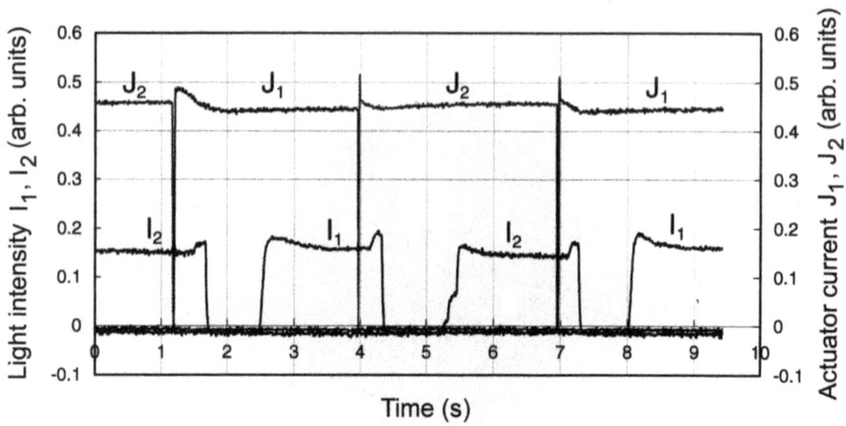

Fig. 6.27. Intensities of transmitted light I_1 and I_2 of output fibers 1 and 2 for alternating control of SMA actuators 1 and 2 with the currents I_1 and I_2 [273]

6.5.3 Conclusions

Optical microswitches are realized by the displacement of a beam block in the optical beam path between two stationary optical fibers, as well as by the active displacement of a launching optical fiber with respect to two stationary output fibers. Table 6.11 summarizes the main specifications of the optical switches, which make the assets and drawbacks of the used switching concepts clear.

Table 6.11. Specifications of optical microswitches

	Optical switch with movable beam block	Optical switch with movable optical fiber
Size	12 x 18 x 4 mm^3	13 x 18 x 1.3 mm^3
Actuation stroke	70 μm	300 μm
Switching force	25 mN	100 mN
Maximum current	200 mA	200 mA
Maximum frequency	1 Hz	1 Hz
Minimum switching time (200 mA, single operation)	100 ms	250 ms
Coupling losses (without blocking of beam)	3 dB	1.2 dB
Crosstalk	-	> 60 dB

For optical switches with a movable beam block, the requirements with respect to actuation stroke and switching force are relatively low. Thus, the specific advantages of SMA linear actuators are not exploited. From the literature, several alternative solutions based on other actuation principles are already known that fulfill the requirements likewise. Concerning the dynamic properties, competing microswitches perform better in most cases [103, 282, 283].

In contrast, the requirements of optical switches with movable multimode fiber are relatively high with respect to force and displacement. In this case, SMA linear actuators offer compact solutions that are difficult to realize with other actuation principles.

Due to relatively long heat transfer times upon cooling of the SMA actuators, only small frequencies on the order of 1 Hz are possible. For pulsed operation with intervals longer than cooling time or for single pulse operation, however, the switching times are determined by the considerably shorter heating time.

The concept of using optical components in between stationary optical fibers is associated with the disadvantage of relatively high coupling losses. Thus, alternative solutions have to be developed for applications with high demands on transmission performance. One possible solution consists of the active displacement of launching fibers with respect to output fibers. The results presented of a 1 x 2 optical microswitch demonstrate that low coupling losses on the order of 1 dB are possible without refractive index matching.

6.6 Microgripper

The increasing integration of mechanical, optical, and fluidic functions in microsystems involves a variety of materials and geometries that cannot be realized by pure monolithic processes. As discussed for the example of the fabrication of SMA microvalves in Sect. 5.3.2.3, only hybrid integration offers the required flexibility in choosing optimum fabrication parameters without affecting the functionality of other system components. Thus, the transition from microelectronic systems to complex microsystems demands an increasing emphasis on hybrid integration technologies.

For the hybrid integration of microcomponents, novel methods for handling and mounting are required. For example, the integration of microlenses in optical microsystems demands precise positioning with respect to the optical beam path and subsequent fixing. Comparable tasks arise for the integration of mechanical or fluidic microcomponents. In these fields, a large need emerges, which is reflected in various research and development projects running presently. Correspondingly, diversified concepts exist for the development of microgrippers, see, e.g., [246, 279, 280, 287–292].

Currently, vacuum grippers are used in the semiconductor industry for the handling of small objects. However, this principle is not suitable for application in vacuum chambers or for the handling of three-dimensional objects. Mechanical methods of gripping based on two or more fingers have the largest flexibility. Macroscopic mechanical grippers cannot simply be scaled to microdimensions. This is due to the generally nonlinear scaling behavior of macroscopic system properties upon miniaturization, as discussed in Sect. 2.1. Furthermore, the fabrication of down scaled systems itself requires the application of a microhandling system. Consequently, novel gripping principles have to be developed that can be realized by microtechnologies. In the following, the development of a mechanical SMA microgripper will be addressed consisting of an SMA linear actuator and an SMA gear actuator.

6.6.1 Requirements of a Mechanical SMA Microgripper

Typical components to be handled in micromechanics, -fluidics, and -optics have dimensions between 20 μm (e.g., wires) and 1000 μm (e.g., ball lenses). Large differences exist with respect to geometry and materials. Thus, flexible solutions are required in order to open up a broad field of applications. The following guidelines should be taken into account:

(1) Reliable performance of gripping and releasing: The adaptation of gripping jaws to the shape and material of the gripping object allows a reliable performance. Adhesion forces $F_{Adhesion}$ have to be controlled upon release, which becomes problematic for smaller gravitational forces of gripping object F_G than $F_{Adhesion}$.

(2) Provision of small loading of gripping object: For decreasing gripping areas, acceptable gripping forces decrease as well in order to keep within critical stress limits upon gripping. If necessary, gripping forces have to be controlled. Generally, gripping forces should be distributed on several points of contact. In the case of thermal actuators, thermal overloading of the gripping object has to be excluded as well.

(3) Observance of relative position of gripping object: A defined position and orientation of the gripping object should be maintained during mounting even in the presence of external forces.

(4) Observance of accuracy of deposition: In microoptics, typical tolerances of final positions are below 1 μm. This requirement can be met by centering the gripping object or by using a position sensor.

(5) System compatibility: Depending on the field of application, the gripper should meet certain requirements with respect to size, cleanroom capability, vacuum capability, compatibility of mechanical and electrical interfaces, etc. If a joining step is peformed after deposition, e.g., by adhesive bonding, soldering, or welding, the gripping concept has to be adapted to the joining technology as well.

As sketched in Fig. 6.28, mechanical grippers generally consist of an actuator and a gearing mechanism, which is connected to gripping jaws. The gearing mechanism transforms the motion of an actuator to a gripping motion. In the present case, the scheme of a shear gripper is shown, where the gripping jaws perform a rotational motion via a joint mechanism. By modifying the gearing mechanism, parallel grippers requiring a larger number of joints may also be realized [288].

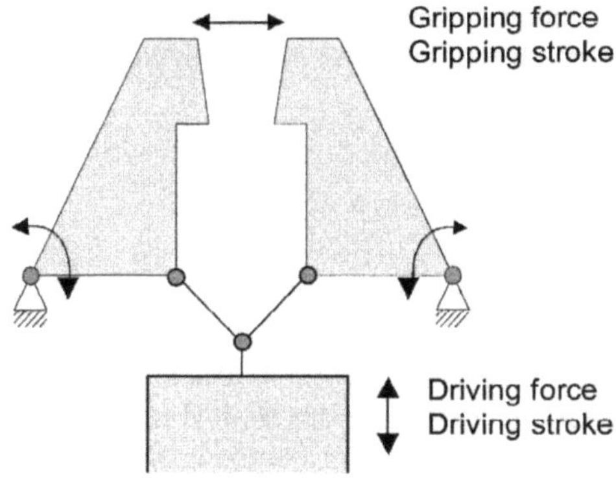

Fig. 6.28. Scheme of a mechanical gripper. The joints of the gearing mechanism are represented as circles

Size requirements particularly govern the gripping concept. Small sizes of gripping objects suggest the use of compact grippers in order to meet guidelines (1)–(5). Assuming an overall size of a few millimiters, all conventional gearing mechanisms consisting of swivel joints and bearings, as well as available hydraulic and pneumatic actuators drop out. Suitable solutions of microsystems technology feature the following characteristics:

- Simple setup: the smaller systems complexity, the higher the achievable degree of miniaturization.
- Gearing mechanism consisting of hinges: Hinges can be easily miniaturized and allow frictionless motion.
- Solid state actuators for actuation: Solid state actuators exhibit a favorable scaling behavior upon miniaturization and allow particularly simple designs. Furthermore, they are free of abrasion and, thus, can be used in cleanrooms without particular measures.

These characteristics are fulfilled by SMA actuators. Thus, it is obvious to implement the scheme in Fig. 6.28 in the design of an SMA actuator. The gearing mechanism and gripping jaws may be designed as hinges and jaws of an SMA thin plate, respectively. For actuation, an SMA linear actuator comes into consideration. Implementation of this idea leads to a completely monolithic gripper, which can be fabricated by a single micromachining step.

As the advantages of SMA actuators gain in particular importance for small dimensions of a few millimeters and below, the development of SMA microgrippers aims at the following specifications:

- Size: a few millimeters
- Potential for further miniaturization
- Minimum gripping force: 10 mN
- Minimum gripping stroke: 200 μm
- Positioning capability (optional)

For soft gripping, a high dynamics is rather undesirable. Insofar, SMA microactuators displaying smooth motions with typical switching times between 10 and 1000 ms suit very well.

6.6.2 Design

In the simplest case, the hinge between the suspension and gripping jaw may be realized by a straight beam, as sketched in Fig. 6.29(a). Assuming a minimum beam width of 50 μm, a displacement of 200 μm, and a maximum strain of 1.5%, the corresponding beam length can be determined to about 1 mm using an elastic beam bending model. A more compact design is achieved by circular beam geometry, see Fig. 6.29(b). This geometry has the additional advantage that external gripping forces cause a torque at the gripping jaws in the direction parallel to the beam and, thus, do not induce any deformations perpendicular to the beam direction.

The hinge between the gripping jaw and linear actuator has to compensate for the horizontal mismatch induced upon circular motion of the gripping jaws, as well as couple the force of the linear actuator to the system. Both functions can also be accomplished by beams of circular geometry without inducing undesirable deformations during motion. Mechanical FEM simulations demonstrate that a minimum beam length on the order of 1 mm has to be maintained for a minimum beam width of 50 μm in order to keep below a maximum strain of 1.5% [272].

The favored geometry resulting from these considerations is shown in Fig. 6.30. Assuming a maximum lateral size of 2 mm perpendicular to the actuation direction of the linear actuator, a maximum radius of beam curvature R of 0.48 mm can be realized. In this case, the beam lengths l_1 and l_2 are 0.8 and 0.9 mm, respectively.

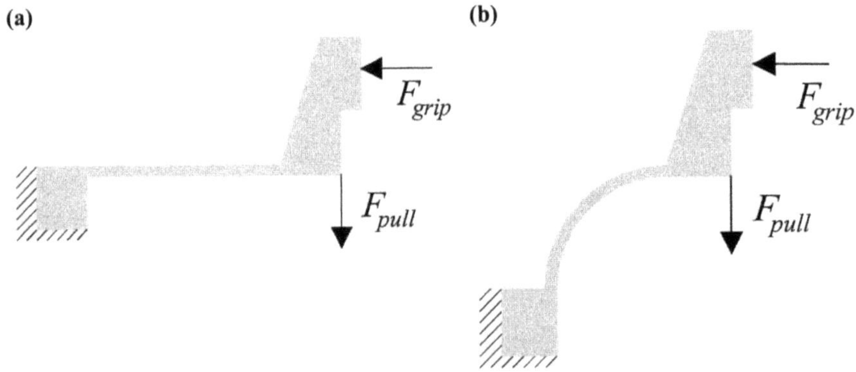

Fig. 6.29. Design of the hinge between suspension and gripping jaw: (a) straight beam, (b) circular beam

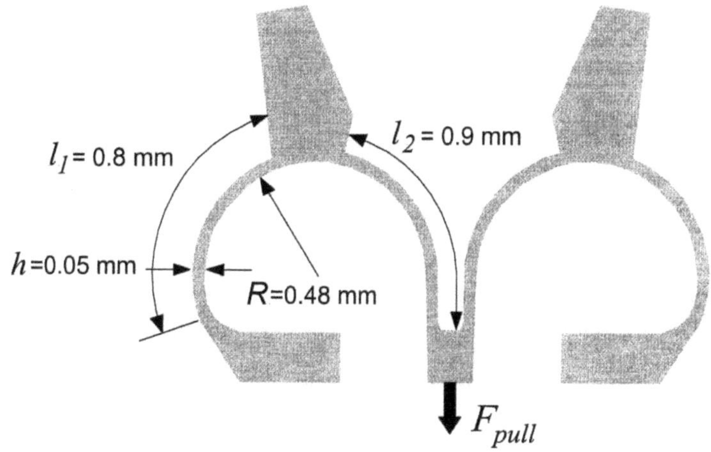

Fig. 6.30. Geometry of an SMA gear actuator consisting of two hinges on both sides of the gripping jaws of width h and lengths l_1 and l_2. R denotes the radius of beam curvature

The gear mechanism presented combines several passive and active functions in the narrowest space:

- mechanical support function
- transfer of linear motion to a gripping motion
- active reset function

Thus, the gear mechanism works as a compact gear actuator. The force of the gear actuator is directed opposite to the force of the linear actuator so that both actuators form an antagonistic pair.

Mechanical FEM simulations reveal that the geometry in Fig. 6.30 shows an inhomogeneous stress profile along the lateral beam surface upon loading by the force F_{pull}. Thus, the shape memory effect is not used in an optimum way. In order to fulfill the mechanical optimization criteria (O1)–(O3), stress optimization is performed using the modified CAO method discussed in Sect. 4.3.1. In this case, the width of the hinges h is used as an optimization parameter. Thereby, a stress limit σ_{limit} of 90 MPa in austenitic condition is assumed.

The SMA microgripper is built up from the optimized SMA gear actuator with gripping jaws and an SMA linear actuator with adjusted force-displacement characteristics. The resulting overall structure is a completely monolithic design, which is shown in Fig. 6.31 for open and closed conditions. For optimization of geometry, the ratio of stress limit to displacement of linear actuator S_l is considered as a further criterion. Thus, optimization of the gear ratio is achieved in addition to stress optimization. The beam width of the optimized SMA gear actuator decreases from 108 μm near the suspension to the minimum value of 50 μm. For the design of the linear actuator, three foldings suffice to generate the required forces and displacements. The optimum profile of the beam width varies between 50 and 114 μm.

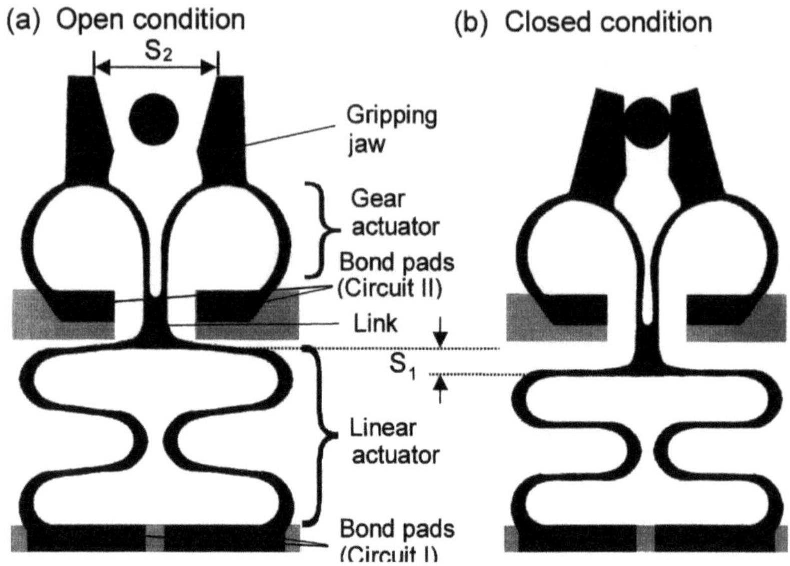

Fig. 6.31. Simulation of motion of a stress-optimized SMA microgripper consisting of a gear actuator, gripping jaws, and a linear actuator [278]

For the operation of SMA microgrippers, a defined predeflection is required, as for linear actuators. For the present simulation, predeflection of the gear actuator with respect to the linear actuator is 150 μm. Fig. 6.31(a) shows the microgripper in open condition. This situation is obtained by the selective transformation of the gear actuator to austenitic condition. In contrast, the selective transformation of the linear actuator results in closed condition, as shown in Fig. 6.31(b) for a circular gripping object.

Simulated profiles of von Mises stress are shown in Fig. 6.32 for an optimized SMA microgripper predeflected by 150 μm. Homogeneous stress profiles along the lateral sidewalls are present in open as well as in closed condition. In open condition, the maximum von Mises stress in the gear actuator reaches 90 MPa. In closed condition, this stress decreases to about 60 MPa. The corresponding maximum von Mises stress in the linear actuator displays somewhat lower values below 60 MPa as long as no external force is present at the gripping jaws. Thus, a sufficient force reserve is provided for gripping.

For testing the thermal optimization criteria (O4)–(O6), thermal FEM simulations are performed for the stationary cases of open and closed conditions. Fig. 6.33 shows simulated temperature profiles on the top surface of an SMA microgripper of 100 μm thickness. The gear and linear actuators have been heated by an electrical power of 22 mW.

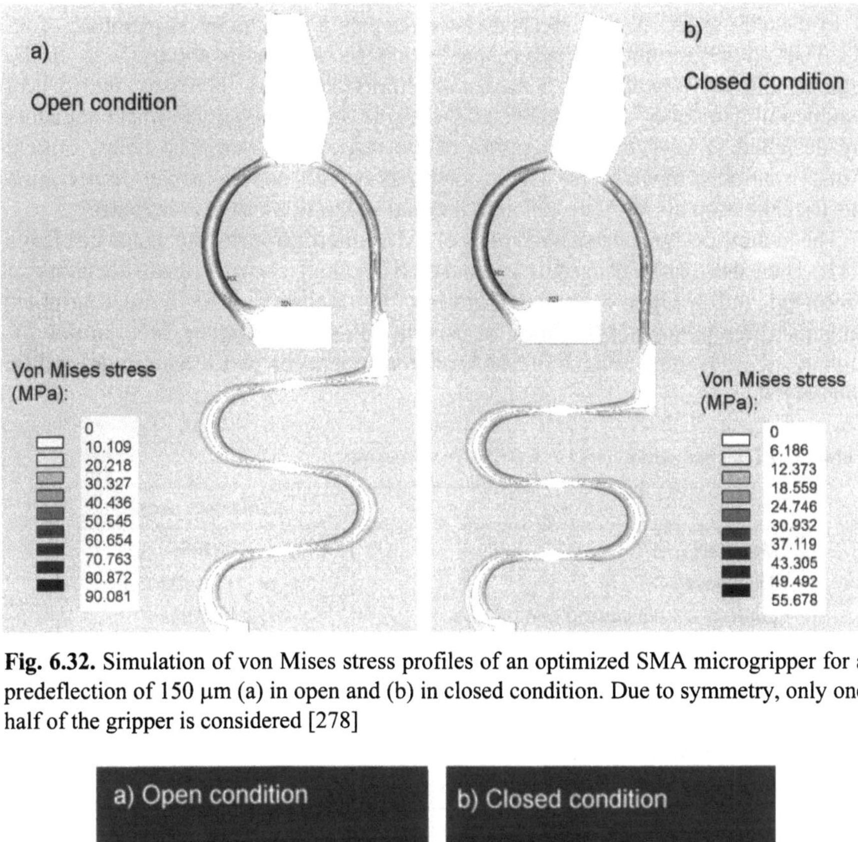

Fig. 6.32. Simulation of von Mises stress profiles of an optimized SMA microgripper for a predeflection of 150 μm (a) in open and (b) in closed condition. Due to symmetry, only one half of the gripper is considered [278]

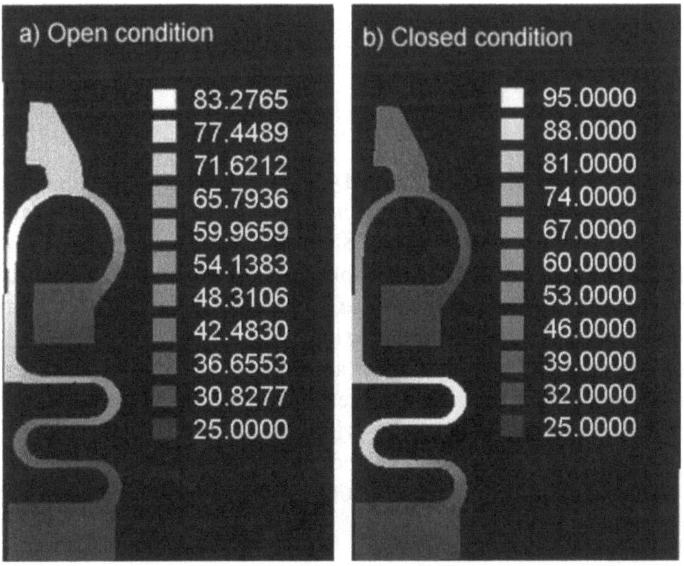

Fig. 6.33. Simulation of temperature profiles of an optimized SMA microgripper (a) in open and (b) in closed condition. Due to symmetry, only one half of the gripper is considered [272]

In closed condition, the linear actuator reaches a maximum temperature of 95 °C. The maximum temperature obtained in open condition is about 85 °C. In the regions that are not directly heated, the temperature also increases due to heat conduction. However, the dimensions of the link between gear and linear actuators are designed to keep the temperature of the respective antagonist below critical transformation temperatures. Thus, compliance with certain power limits upon electrical heating allows the sufficient thermal separation of the antagonists.

The major design parameters of the SMA microgrippers are listed in Table 6.13. The base material used is a binary NiTi alloy, as with linear actuators of variants 1 and 3. Upon stress optimization, the thickness of SMA microgrippers remains a free parameter that may be used to adjust the gripping force. In the following, two different designs of SMA microgrippers for two force ranges will be considered.

Table 6.12. Design parameters of two SMA microgrippers

	Variants 1 and 2
Material	NiTi
Thickness	100 μm (1) and 230 μm (2)
Lateral size (including bond pads)	2.1 x 3.9 mm^2
Minimum width of SMA gear actuator	55 μm
Maximum width of SMA gear actuator	108 μm
Number of foldings of SMA linear actuator	3
Minimum width of SMA linear actuator	50 μm
Maximum width of SMA linear actuator	114 μm

6.6.3 Fabrication

Due to the monolithic design of the SMA microgrippers, fabrication can be performed in a single process step by laser cutting a NiTi plate, as for linear actuators of variants 3 and 4. Fig. 6.34 shows an SMA microgripper fabricated from a NiTi plate of 100 μm thickness. Due to technological reasons, the bond pads at the rear end are interconnected by a small bridge, which has to be intersected after being mounted on a substrate, e.g., by laser cutting.

Before mounting, the gear and linear actuators are predeflected with respect to each other using a precision table and then fixed on a substrate by adhesive bonding. Electrical contacts are made by wire bonding or laser welding. Fig. 6.35 shows an SMA microgripper laser-welded on a ceramic substrate with Ni conductor paths. Below the movable microstructures, the substrate has been removed to enable frictionless motion.

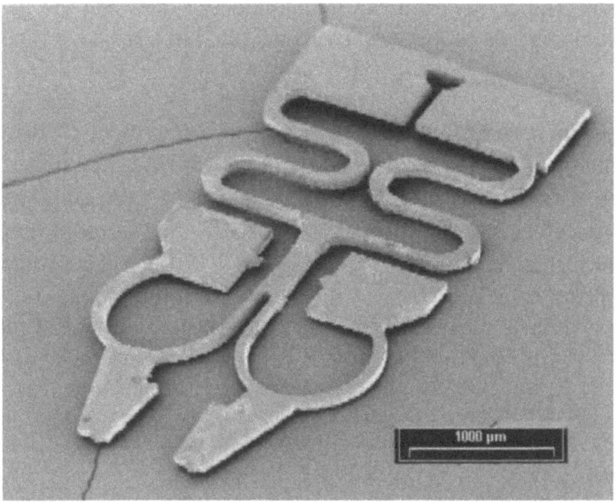

Fig. 6.34. SMA microgripper of variant 1 after laser micromachining

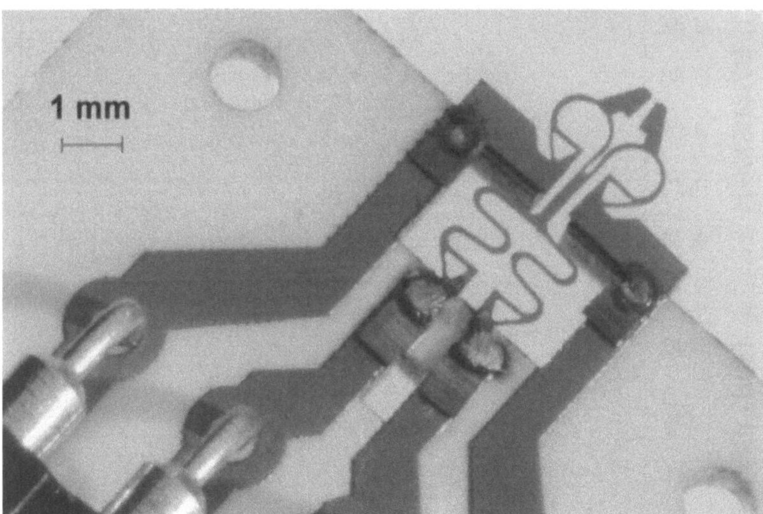

Fig. 6.35. SMA microgripper of variant 1 after mounting on a substrate

6.6.4 Performance Characteristics

At first, the gear and linear actuators are investigated separately to obtain a detailed understanding of the overall physical properties of the SMA microgripper. Force-displacement characteristics are determined by orienting the actuators vertically and attaching calibrated weights to the coupling element, which, in case of the SMA microgripper, serves as an interconnection between the gear and linear actuators. The functional properties are investigated in stationary condition as well as a function of time.

Stationary performance: Fig. 6.36 shows characteristics of electrical resistance and displacement of an SMA linear actuator of 100 μm thickness. The external load has been varied between 26 and 42 mN. In austenitic condition, the corresponding displacements range from about 14 to 28 μm. At room temperature, the corresponding displacements are between 70 and 110 μm. These results allow an estimate of spring constants. In austenitic condition, the spring constant is about 1,700 N/m. At room temperature, the effective spring constant at maximum displacement of 110 μm is about 380 N/m.

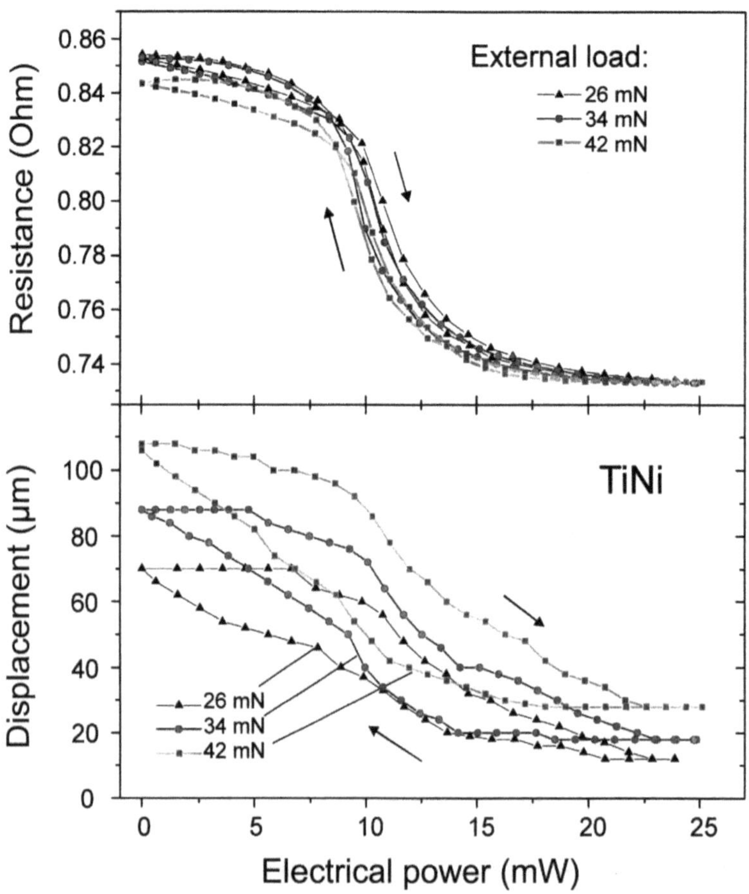

Fig. 6.36. Characteristics of electrical resistance and displacement of an SMA linear actuator of 100 μm thickness [278]

The electrical resistance shows the characteristic features of R-phase transformation, as for linear actuators of variants 1 and 3. The increased hysteresis width of displacement characteristics is caused by SIM formation occurring locally in the regions of maximum stress along the sidewalls. These regions mainly affect the displacements, but have almost no influence on the integral resistance of the components. The power requirement displacing the linear actuator is about 25 mW.

Fig. 6.37 shows the typical characteristics of the electrical resistance and stroke of gripping jaws of an SMA gear actuator of 230 μm thickness. At room temperature, the gripping jaws are completely closed for external loads above about 80 mN. For this load, the gripping jaws exhibit a maximum stroke of about 300 μm. The corresponding gear ratio given by the ratio of strokes of the linear and gear actuators S_1/S_2 results to 2.3. For smaller loads, the gripping jaws are not completely closed. Higher loads than 80 mN cause a decrease in the maximum displacement and stroke of the gripping jaws. The gripping force can be estimated from the gear ratio to be 35 mN. The pronounced hysteresis of the displacement characteristic is only partly due to SIM formation in the regions of maximum stress. Furthermore, inhomogeneous temperature profiles in the gear actuator lead to an increase of power requirement in order to obtain complete phase transformation.

SMA microgrippers of 100 μm thickness display similar characteristics. In this case, the linear actuator has to exert a force of about 35 mN to completely close the gripping jaws. The gripping force is estimated to be 17 mN. The spring constants of the gear and linear actuators are summarized in Table 6.14 for both variants of microgrippers. Based on the spring constants, displacements and strokes can be estimated as a function of predeflection, as described in Sect. 6.4.1.

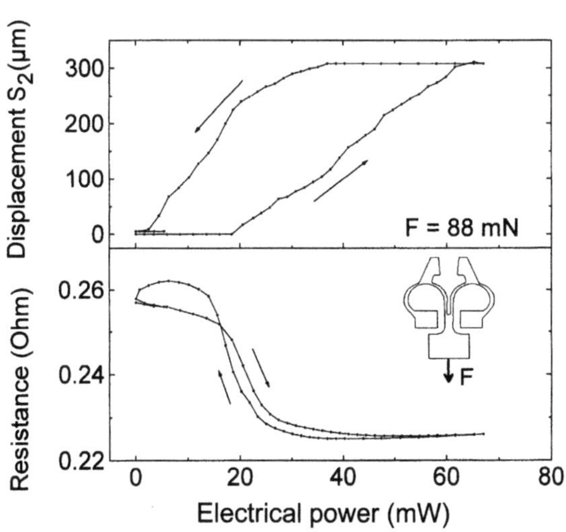

Fig. 6.37. Electrical resistance and displacement of an SMA gear actuator of 230 μm thickness [276]

Table 6.13. Spring constants of the gear and linear actuators of variants 1 and 2 in austenitic condition and the corresponding effective spring constants for the case of maximum displacement at room temperature (R-Phase and stress-induced martensite SIM)

		Variant 1	Variant 2
Gear actuator:	Austenite	1,650 N/m	4,300 N/m
	R-phase / SIM	350 N/m	600 N/m
Linear actuator:	Austenite	1,700 N/m	4,450 N/m
	R-phase / SIM	380 N/m	580 N/m

Dynamic performance: The response time of the linear actuator upon heating can be determined by time-resolved electrical resistance measurements. During heating, the electrical resistance decreases and finally reaches a minimum, which coincides well with the final position of the SMA actuator, as shown in Fig. 6.36. The response times of an SMA linear actuator of 100 µm thickness are shown in Fig. 6.38 as a function of electrical power. The inset shows a typical time-resolved electrical resistance characteristic for a heating power of 17 mW.

The response times decrease inversely proportional to P_{el}, as expected for adiabatic heating. At a heating power of 22 mW, the response time decreases to 32 ms. The corresponding time constants upon cooling are on the order of 300 ms without active cooling. Due to the antagonistic mechanism, however, the switching times for both moving directions are determined by the response times upon heating. The cooling times determine the maximum possible frequency for passing through complete actuation cycles. In the present case, this frequency limit is about 3 Hz.

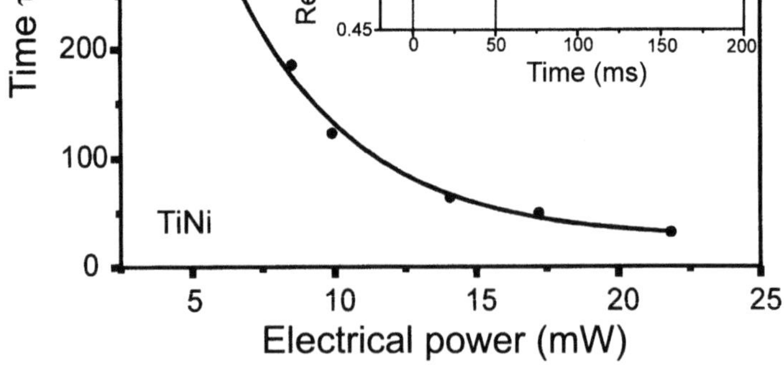

Fig. 6.38. Response times of a linear actuator of 100 µm thickness [275]

SMA microgrippers of 230 μm thickness exhibit a response time of 50 ms for a heating power of 65 mW. The corresponding time constant upon cooling is on the order of 400 ms.

Temperature profiles along the surface of SMA microgrippers have been investigated by infrared microscopy as a function of time upon heating and cooling. In order to obtain a homogeneous coefficient of thermal emission equal to 1, the surface is coated with a thin layer of graphite. The maximum spatial resolution of the optical setup used is about 25 μm, and the maximum temporal resolution about 100 ms.

Fig. 6.39(a) shows the temperature profile of an SMA microgripper of 100 μm thickness obtained upon heating the gear actuator by a power of 20 mW. The induced temperatures of hinges and gripping jaws are in the range between 50 and 85 °C. In this case, the gear actuator is in austenitic condition. In the region of the coupling element between the gear and linear actuators the temperature decreases below 40 °C. Thus, the linear actuator remains in R-phase condition.

After subsequent cooling of the gear actuator for 150 ms, the temperature profile shown in Fig. 6.39(b) is observed. If the heating power of the linear actuator is now switched on, the temperature profile shown in Fig. 6.39(c) is generated after another 150 ms. Within this time period, a stationary temperature profile has attuned in the linear actuator. In this case, temperatures up to 100 °C are reached. At the same time, cooling of the gear actuator is almost complete. Thus, the gear actuator is again in pure R-phase condition.

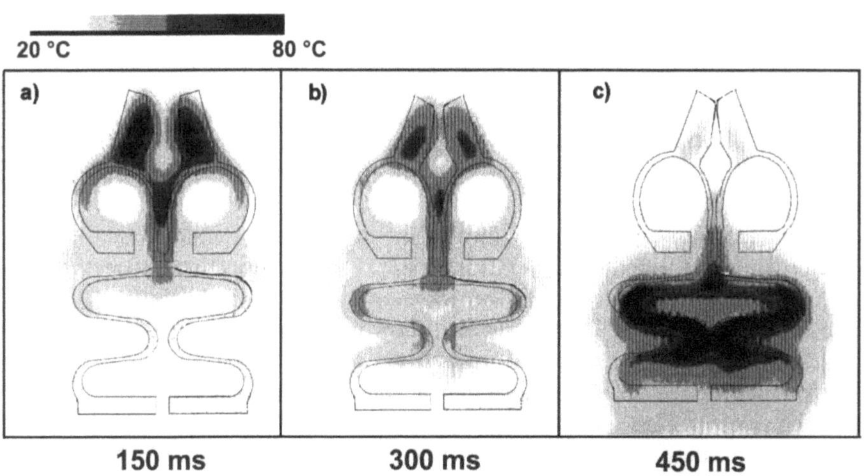

Fig. 6.39. Sequence of temperature profiles along the surface of an SMA microgripper of variant 1 determined by infrared microscopy (a) after heating the gear actuator for 150 ms, (b) after a cooling period of 150 ms, and (c) after heating the linear actuator for 150 ms. Regions at room temperature are indicated in white, regions above 100 °C in black [275]

These investigations confirm the result of thermal FEM simulations discussed in Sect 6.5.2 that the link in between the antagonists provides a sufficient thermal isolation. Thus, selective control of the antagonists is ensured. In closed condition, the temperature of the gripping jaws is near room temperature, so that no thermal loading of the gripping object occurs. Furthermore, the compliance in R-phase condition gives rise to a certain softness upon gripping.

6.6.5 Positioning

Many applications in robotics require a precise knowledge of intermediate positions between the extremes of "gripper open" and "gripper closed." This facilitates the manipulation of objects with complex shapes by the adaptation of gripping position, it also enables the control of gripping force by the difference of the geometrical dimensions of the gripping object and set value for the gripping position.

The hysteresis effects of the displacement characteristics, shown, e.g., in Figs. 6.36 and 6.37, raise the problem of a suitable control method for positioning. Recent investigations on certain NiTi wires revealed an almost linear relationship between length change and electrical resistance [201], which enables smart actuation by using the sensing capability of the wire itself. However, this feature strongly depends on the thermomechanical treatment of SMA material and the load-dependent distribution of stress. In particular, SMA microactuators display nonuniform stress and temperature profiles due to varying lateral width profile. Consequently, nonuniform distributions of R-phase and martensitic phase arise, which affect the electrical and mechanical characteristics in a different way. In the present case, SIM occurs in the regions of maximum stress along the sidewalls. This is associated with a hysteresis broadening in displacement characteristics, while the corresponding electrical resistance characteristics remain almost unaffected. Thus, the corresponding displacement-electrical resistance characteristics also display large hysteresis widths, which cannot be used for smart control of the SMA microgrippers.

Another option for position control is the integration of an additional position sensor combined with a feedback control loop. Preliminary investigations showed that position sensing with a light barrier is a suitable method that is relatively insensitive to temperature changes and adjustment tolerances during fabrication. The light barrier is used to detect the motion of the SMA linear actuator in order not to obstruct the workspace of the SMA gear actuator and gripping jaws. The SMA linear actuator performs a linear displacement in between the end positions of completely open and closed gripping jaws, which directly reflects the gripping position due to the known gear ratio.

By a small change of gripper design, an optical slit is integrated in between both antagonists. The slit performs the same motion as the SMA linear actuator and, thus, can be used to interrupt or deblock the optical beam path of a light barrier arranged perpendicular to the motion direction. Fig. 6.40 shows an SMA microgripper of 230 μm thickness with an integrated slit.

Fig. 6.40. SMA microgripper of 230 µm thickness with an integrated slit in between the gear and linear actuators

By integrating the optical position sensor, an optomechanical gripping system is formed, which is sketched in Fig. 6.41. The SMA gripping device is mounted in between a lower and an upper substrate with integrated LED and photosensor, respectively, which are aligned with respect to the optical slit. The lower substrate contains pads for mechanical and electrical interconnections of the SMA microgripper. Lower and upper substrates are connected to each other via a spacer.

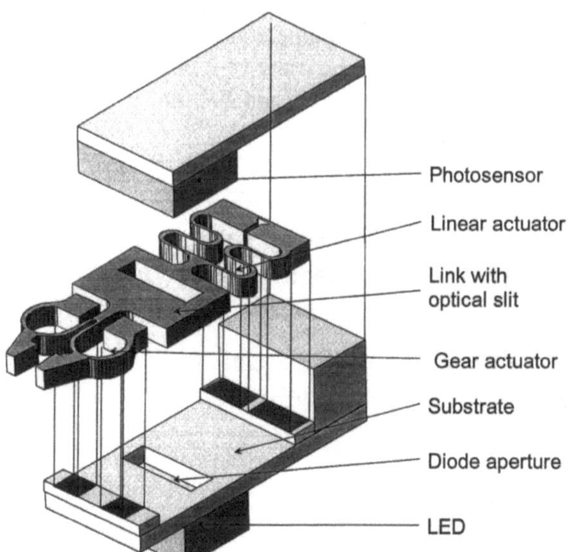

Fig. 6.41. Scheme of an optomechanical gripping system consisting of an SMA microgripper with an integrated optical position sensor

By measuring the optical transmission signal, the actual position of the optical slit and, thus, the actual gap in between the gripping jaws can be determined. Fig. 6.42 shows a calibration characteristic of the optical transmission signal. The displacement of the optical slit has been adjusted by a precision table. Within the range of gripper displacements, a linear behavior is present. In the regions of maximum and minimum transmissions, S-like transitions are observed, which are not shown here. In the linear range, a spatial resolution of 1 μm corresponds to a signal change of 0.7%. For a measurement error of about 0.5%, the corresponding resolution of displacement is 0.7 μm.

For test purposes, the microgripper system is controlled by a PC using a data aquisition board. The optical transmission intensity detected by the photosensor is processed by a Proportional-Integral (PI) control algorithm. The PI-control parameters are determined by an empirical routine in order to obtain an optimum control velocity and accuracy. A typical step-response of control and sensor signals is shown in Fig. 6.43. The algorithm generates negative and positive control voltages in order to discriminate between the two antagonists. The maximum power to drive the actuators is set to 80 mW, which corresponds to a maximum control signal of 10 V. The control frequency is 125 Hz.

In this example, the initial position is maintained by an average control signal of 2.5 V. After changing the set value for transmission intensity from 0.65 to 0.5, the SMA linear actuator is driven with maximum power for about 120 ms. The transmission signal reflects the corresponding change in actual position, which takes about 140 ms. The difference between set and actual values corresponds to a positioning accuracy of about 0.5%, which is achieved in the range of 10 to 90% of maximum gripping stroke. In this range, the positioning accuracy of the optical slit is determined to be 0.7 μm. Assuming a gear ratio of 2.3, the resulting resolution of the gripping position is on the order of 2 μm, which is sufficient for gripping objects with dimensions between 20 and 200 μm.

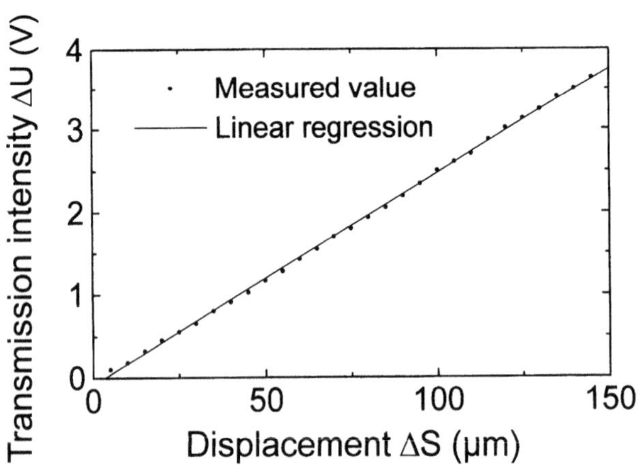

Fig. 6.42. Calibration characteristic of the optical transmission signal of an SMA microgripper of 230 μm thickness [293]

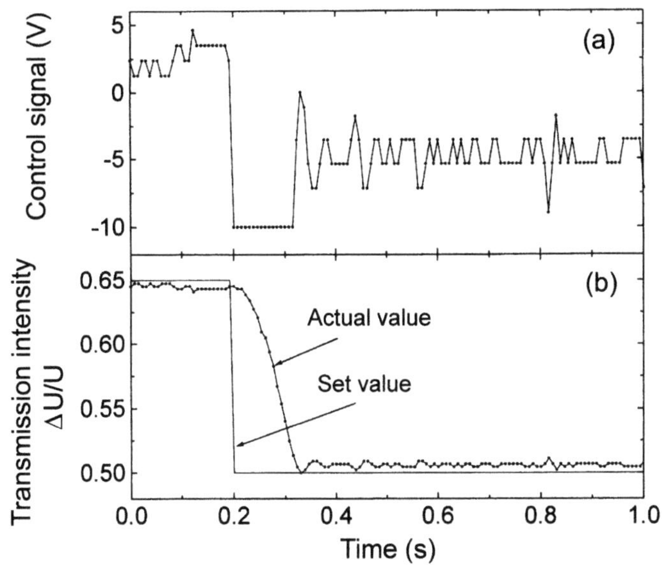

Fig. 6.43. Step-response of the signals of PI control (a) and positioning sensor (b) from an SMA microgripper of 230 μm thickness [293]

Fig. 6.44 shows the response to an arbitrary sequence of set values. Close to the end positions, the positioning time increases considerably to about 0.5–1 s, depending on the maximum power for gripper control. For most applications, however, these delays represent no constraint, as completely closed and open positions normally serve as standby positions.

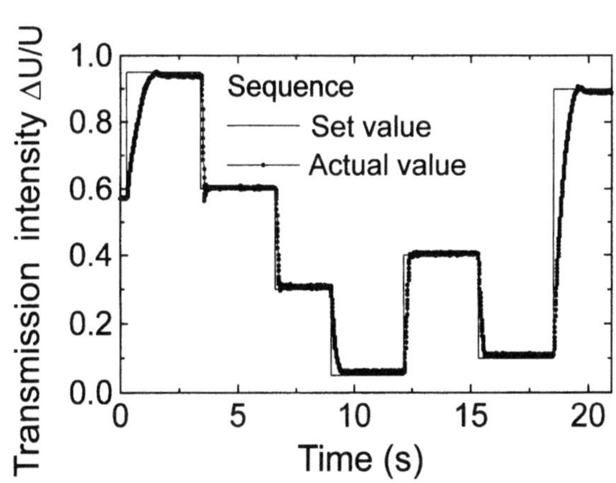

Fig. 6.44. Response of an SMA microgripper of 230 μm thickness to an arbitrary positioning sequence [293]

6.6.6 Conclusions

By coupling an SMA linear actuator and an SMA gear actuator, a completely monolithic mechanical SMA microgripper is designed that can be fabricated from one piece of material in a single fabrication step. The simultaneous performance of various functions, such as structural, electrical conduction, gearing, antagonistic actuation and gripping functions, by a single compact component makes the SMA microgripper a paradigm of microactuation. By the additional integration of a light barrier, a compact optomechanical gripping system is formed that fulfills high demands of smallness, work output, and positioning accuracy.

The main specifications of the presented microgripper variants are summarized in Table 6.15. The size of the SMA gripping system for positioning is about 2.5 x 8 x 9 mm^3, including the light barrier. By miniaturizing the light barrier, particularly the height can be reduced considerably. The switching and cycling times are in the ranges of 30–50 ms and 300–500 ms, respectively. In the range of between 10 and 90% of maximum gripping stroke, the positioning time is 140 ms. Thus, the dynamic performance of the SMA microgripper is also attractive for various applications such as handling microparts of complex shape in restricted environments, vacuum environments, or clean rooms. For comparison, typical time constants for position control by digital imaging are on the order of 300 ms [294]. So far, the SMA microgripper has been tested in assembly robotics for optical microcomponents [295].

Table 6.14. Specifications of microgripper variants 1 and 2. The indicated lateral size refers to SMA microgrippers without position sensor

	Variant 1	Variant 2
Thickness	100 μm	230 μm
Lateral size	2.1 x 3.9 mm^2	2.1 x 3.9 mm^2
Maximum gripping force	15 mN	35 mN
Maximum gripping stroke	300 μm	300 μm
Positioning accuracy	2 μm	2 μm
Minimum power requirement	20 mW	60 mW
Minimum switching time	30 ms	50 ms
Minimum cycling time	330 ms	450 ms

One outstanding feature of this SMA microgripper is the multifunctional compact design, which is easily scalable to smaller or larger dimensions. Thus, the range of applications can be simply extended to gripping objects of various sizes and to various gripping forces. The mechanical gripping principle and the easily changeable shape of the gripping jaws allow a high flexibility with respect to dimensions and shape of the gripping object. Due to the metallic properties of the SMA microgripper, the electrical potential of the gripping jaws can be grounded or adapted to the gripping object in order to minimize charging or adhesion effects.

Table 6.16 contains specifications of other microgrippers based on an SMA thin film composite (see Sect. 4.5.1, [246]) and a thermal bimorph [289], both fabricated monolithically using a combination of thin film deposition and silicon technology. For comparison, the specifications of a miniature gripper driven by a piezo actuator are listed as well [291]. In most cases, miniature mechanical grippers make use of a passive bending mechanism driven by an actuator, which is fabricated separately and, thus, must be interconnected to the bending mechanism. Typical examples are SMA wire actuators [288] or piezoelectric actuators [287, 290, 291]. The overall dimensions of these types of grippers are in the centimeter range.

While the maximum gripping force of the microgripper based on bimetallic principle is about two orders of magnitude lower compared to the presented SMA microgripper, the microgripper based on an SMA thin film composite exhibits a comparatively favorable combination of gripping force and stroke. However, the specifications achieved with respect to switching time and power requirement come off rather badly. This result is due to the thermal coupling between the Si and NiTiCu layers, which prevents selective heating of the NiTiCu layer. Furthermore, this type of microgripper requires a large fabrication effort, which includes, e.g., precision sawing [246].

Table 6.15. Specifications of alternative microgrippers and a miniature gripper based on piezoelectric effect

	Shape memory effect	Bimetal effect	Piezoelectric effect
Reference	[246]	[289]	[291]
Lateral size (mm^2)	0.38 x 1	3.5 x 1	15 x 22
Maximum gripping force (mN)	13	0.25	15
Maximum gripping stroke (μm)	110	140	50
Power requirement (mW)	320*	80	-**
Switching time (ms)	500	10	-**

* Indirect heating by external heating element

** No data published

6.7 Outlook

In this chapter, the concept of hybrid integration of micromachined SMA plates with optimized lateral shape has been followed up for the realization of SMA linear actuators and related SMA microactuators. This concept allows the integration of variegated functions in a single component of monolithic structure with compact size, while fabrication remains remarkably simple. Micromaching, as well as hybrid integration, can be performed in parallel, allowing an extremely cost-effective fabrication. Due to high energy densities, large forces and displacements are achieved. For overall dimensions of a few millimeters, typical time constants are on the order of 100 ms and below.

Micromotion systems like the linear actuators presented, optical microswitches, and microgrippers are of interest for various application fields such as:

- microrobotics
- automation technology
- medical technology
- safety engineering

As forecasted in the fifties by the physicist and nobel laureate R.P. Feynman, the development of microsystems necessitates the fabrication of microrobots [296]. This applies particularly to hybrid microsystems. In recent years, the demand on ultrafine positioning systems has undergone a significant upturn, which is caused by the increasing trend of miniaturization in numerous fields [133]. Currently, micro- and nanorobots are developed for the ultraprecise handling of microobjects working, e.g., in a scanning electron microscope [289]. Medical technologies lead the way in the development of compact motion systems, particularly in the field of minimal invasive therapy [297]. Several projects in the field of safety engineering address the development of complex locking mechanisms based on micromotion systems [298, 299].

Further objectives for opening up a broader field of applications for SMA microactuators are:

- SMA linear actuators for operation temperatures above 100 °C: While the fabrication of thin films based on NiTiPd and NiTiHf with suffficiently high transformation temperatures has already been demonstrated successfully, the fabrication of thin plates based on these alloys in the thickness range of 100 μm is still a large challenge.
- Improvement of positioning accuracy: The development of improved control strategies is required in order to further minimize the effects of fatigue and hysteresis. Instead of using PI control, the implementation of digital control strategies based, e.g., on inverse models or neuronal networks, is conceivable.
- Increased portfolio of motion mechanisms: The development of novel motion systems requires the exploitation of additional mechanisms for gearing, guidance, bistability, etc. and the implementation of these mechanisms in the design of SMA microactuators. This may lead, for instance, to novel SMA microgrippers with parallel guided gripping jaws or with bistable gripping mechanisms.
- Free configurable motion systems: By combining several SMA microactuators, complex SMA motion systems may be realized. The concept of SMA linear actuators may be extended, for instance, to coupled SMA linear actuators for two or three degrees of freedom.
- Large-scale production: For cost-effective micromachining of SMA plates in large quantities, laser cutting is less suitable. Therefore, the development of alternative technologies such as etching or punching is of interest. Furthermore, suitable mounting and interconnection technologies have to be developed for batch processing.

7 Summary

Now, you might say, "Who should do this and why should they do it?" Well, I pointed out a few of the economic applications, but I know that the reason that you would do it might be just for fun.

–R.P. Feynman in *There's Plenty of Room at the Bottom*

Various SMA microactuators have been developed since the introduction of SMAs in microsystems technology at the beginning of the nineties. The SMA microvalves and linear actuators discussed in detail in this book show promising specifications reflecting the advances in the fields of material development, modeling, actuator design, micromachining, and system integration. These microactuators are not well engineered products but prototypes that have been fabricated in small quantities. Generally speaking, SMA microactuators are in an early development stage. The progress made so far already demonstrates, however, that SMA microactuators have a large potential for innovation and may contribute significantly to future developments in microsystems technology.

The motivation for using SMAs in microsystems technology is based on the high energy densities and multifunctional properties, which enable simple designs with a large potential for miniaturization. As the disadvantages of low thermodynamic efficiency and low heat transfer rates become less important for increasing miniaturization, advantages with respect to alternative actuation principles are expected particularly for small dimensions.

Material development: Previous material developments lead to novel SMA plates and sputtered SMA thin films, which are the base materials for fabricating SMA microactuators within the framework of the developments presented. Of particular importance are rolled NiTi plates of thickness range 100 μm, whose further development to NiTi foils with thicknesses on the order of 10-20 μm has been primarily motivated by some of the demonstrators presented. Sputtered NiTiPd thin films are also of special interest due to their large flexibility in adjusting phase transformation temperatures.

The increase in application diversity of SMA microactuators requires further material developments. These include, for instance, SMA plates with phase transformation temperatures above 100 °C or SMA thin film systems with adjustable hysteresis or extended transducing properties like ferromagnetic SMAs.

Modeling: By extension of the one-dimensional Tanaka model, a macromodel for the physical properties of three-dimensional SMA components is described. Implementation of the model in a FEM routine for coupled multilevel simulation enables a quantitative description of the electrical, thermal, and mechanical properties of SMA microactuators. This procedure is particularly important for microsystems, as a strong coupling of neighboring components occurs and lateral micromachining allows a large variety of structures.

Macroscopic models only have a limited significance due to the use of averaged material parameters. Further developments aim at three-dimensional FEM models, which reflect the physical processes with sufficient accuracy, as well as contain an optimum degree of simplification in order to achieve adequate simulation times.

Design: Mechanical and thermal optimization criteria are introduced and implemented in a procedure for the optimized design of SMA microactuators. For mechanical optimization, the principle of stress-optimized growth based on the computer-aided optimization method is applied. Thermal optimization is accomplished by strategic arrangement of heat sources and sinks. By design optimization of the SMA linear actuators, an improvement of work density by more than a factor of 6 is achieved. Thermally optimized NiTi microvalves show operation frequencies on the order of 15 Hz.

The relatively complex material properties of SMAs are one reason why the design of SMA actuators is still a domain for specialists. Thus, the development of computer-based design tools and data bases of SMA materials is essential for a broader application.

Micromachining: The procedure presented in this book is based on the lateral micromachining of SMA plates, foils, or thin films. For this purpose, etching processes are pursued particularly for material thicknesses below 50 μm. Laser cutting is favored for plates above 50 μm thickness due to more favorable aspect ratios.

The available technologies must be further optimized with respect to dimensional accuracy, steepness of walls, surface quality, etc. For cost-effective realization of large quantities, laser cutting of SMA plates is only contingently suitable. In this case, the development of alternative technologies such as etching or punching is of large interest.

System integration: A dedicated approach to the fabrication of SMA microactuators is presented based on hybrid integration of micromachined SMA components. Thus, trouble-free handling and precise mounting of sensitive microstructures are feasible. Compared to monolithic integration technologies, a considerably greater flexibility in the choice of process parameters and material combinations results. Thus, SMA microactuators can be easily combined with polymer housings of limited temperature stability, which is of particular importance for medical applications like SMA microvalves. Fabrication of valve components and hybrid integration can be performed in parallel to achieve cost effectiveness.

The integration of many functions in a single component results in extremely compact monolithic designs of SMA microactuators, which enable revolutionarily simple and cost-effective fabrication using a single process step. The presented SMA microgrippers, for instance, comprise various functions like mechanical support, electrical conductance, gearing, antagonistic actuation, and gripping in a single component of a few millimeters in size.

The technologies of sacrificial layers, separation, mounting, and bonding have to be further improved with respect to material variety, dimensional accuracy, reproducibility, etc. Cost-effective fabrication in parallel requires the extension of hybrid integration technologies to batch mode.

Some of these SMA microactuators are already competitive components. Of particular importance are, for instance, normally open SMA microvalves with SMA microactuators of 20 μm thickness, presented in Sect. 5.3, which are counted among the smallest microvalves and are still able to control pressure differences and flows comparable to alternative microvalves of larger size. These microvalves are fabricated in a modular way and, thus, may be combined flexibly with other fluidic microcomponents on a common base plate in order to realize fluidic microsystems in the narrowest spaces. Another example are SMA microgrippers presented in Sect. 6.6, which exhibit particularly large gripping forces and strokes.

SMA microactuators are base components that may be used in complex microsystems. In this book, fluidic, mechanical, and optical microsystems are discussed as potential fields of application. These fields represent, however, only a section of the whole breadth of microsystems technology, which is currently in a highly dynamic development process. The plenitude of applications provides incentives for further research and development activities on SMA microactuators. The prospects are new, emerging markets of microsystems applications.

References

[1] H. Janocha, Aktoren, Springer, Berlin, Heidelberg, (1992)

[2] M. Tabib-Azar, Microactuators, Kluwer Academic Publishers, (1998)

[3] E. Bassous, H.H. Taub, and L. Kuhn, Ink jet printing nozzle arrays etched in silicon, Appl. Phys. Lett. 31 (1977), pp. 135–137

[4] S.C. Terry, J.H. Jerman, and J.B. Angell, A gas chromatographic air analyzer fabricated on a silicon wafer, IEEE Trans. Elec. Dev. ED-26 (1979), pp. 1880–1886

[5] N. Ünal and R. Wechsung, Reality of MST-market on the example of ink-jet print head, MST News 1 (1998), pp. 32–34

[6] Data sheet Dr. Fritz Faulhaber GmbH & Co. KG, Series BL 1900

[7] D.J. Jendritza, Technischer Einsatz Neuer Aktoren, Expert, Renningen-Malmsheim, (1995)

[8] J. van Humbeeck, D. Reynaerts, and R. Stalmans, Shape memory alloys: functional and smart, Proc. Actuator 94, Bremen, Germany (1994), pp. 312–316

[9] J. van Humbeeck, On the adaptivity of shape memory alloys for use in adaptive materials, Proc. ICIM 96, Lyon, France (1996), pp. 442–451

[10] A.D. Johnson and J. Krämer, State-of-the-art of shape memory actuators, Proc. Actuator 98, Bremen, Germany (1998), pp. 457–460

[11] D. Honma, Y. Miwa, and N. Iguchi, Micro robots and micro mechanisms using shape memory alloy, in: Integrated Micro-Motion Systems, Ed. F. Harashima, Elsevier Amsterdam (1990), pp. 375–392

[12] J.A. Walker, K.J. Gabriel, and M. Mehregany, Thin-film processing of NiTi shape memory alloy, Sensors and Actuators A, 21–23 (1990), pp. 243–246

[13] B. Walles, L. Chang, and D.S. Grummon, Residual stress, adhesion and crystallization of ion-sputtered and IBED processed NiTi films, Mater. Res. Soc. Symp. Proc., Vol. 246 (1991), pp. 349–354

[14] A.D. Johnson, Vacuum-deposited TiNi shape memory film: characterization and applications in microdevices, J. Micromech. Microeng. 1 (1991), pp. 34–41

[15] K. Kuribayashi, S. Shimizu, T. Nishinohara, T. Taniguchi, M. Yoshtake, and S. Ogawa, Trial fabrication of micron sized arm using reversible TiNi alloy thin film actuators, Proc. IEEE/RSJ 93, Yokohama, Japan (1993), pp. 1697–1702

[16] S. Miyazaki and K. Nomura, Development of perfect shape memory effect in sputter-deposited Ti-Ni thin films, Proc. MEMS 94, Oiso, Japan, IEEE Catalog No. 94CH3404-1, (1994), pp. 176–181

[17] H. Holleck, S. Kirchner, E. Quandt, and P. Schloßmacher, Preparation and characterization of NiTi thin films, Proc. Actuator 94, Bremen, Germany (1994), pp. 361–364

[18] L. Hou, T.J. Pence, and D.S. Grummon, Structure and thermal stability in Ti-Ni thin films sputtered at elevated temperature on inorganic and polymeric substrates, Mat. Res. Soc. Symp. Proc. Vol. 360, (1995), pp. 369–374

[19] K. Kitamura, S. Miyazaki, and M. Kohl, Anisotropy of shape memory behavior of rolled TiNi thin plates, Proc. Actuator 96, Bremen, Germany, Ed. H. Borgmann (1996), pp. 401–404

[20] M. Matsumoto, T. Suzuki, K. Yamauchi, and S. Nohara, Characterization of TiNi thin film, Proc. China-Japan Bilateral Symp. on Shape Memory Alloys, Eds. C. Youyi and K. Otsuka, Hangzhou, China (1997), pp. 108–113

[21] S. Fatikow and U. Rembold, Microsystem technology and microrobotics, Springer, Berlin (1997)

[22] S. Hirose, K. Ikuta, and Y. Umetani, The development of shape memory alloy actuator, J. of the Robotic Society of Japan, Vol. 4, No. 2 (1986), pp. 89–100

[23] M. Kohl, Fluidic actuation by electrorheological microdevices, Mechatronics 10 (2000), pp. 583–594

[24] A. Wallash and L. Levit, Electrical breakdown and ESD phenomena for devices with nanometer-to-micron gaps, Proc. Int. Conf. on Micromachining and Microfabrication – Reliability, Testing, and Characterization of MEMS/MOEMS II, San Jose, California, USA, SPIE Vol. 4980, (2003), pp. 87–96

[25] B. Rogge, Entwicklung, Optimierung und Test von elektromagnetischen Linearaktoren basierend auf LIGA-Mikrotechnik, thesis, University of Karlsruhe (1996)

[26] W.S.N. Trimmer, Microrobots and micromechanical systems, Sensors and Actuators 19 (1989), pp. 267–287

[27] P. Ruther, Entwicklung eines LIGA-Mikrosystems zur Messung mechanischer Eigenschaften von Mikroproben, thesis, University of Karlsruhe (1997)

[28] P. Ruther, W. Bacher, K. Feit, and W. Menz, LIGA microtesting system with integrated strain gauges for force measurement, Proc. MEMS 97, Nagoya, Japan, IEEE Catalog No. 97CH36021, (1997), pp. 541–545

[29] M. Tabib-Azar and J. S. Leane, Direct optical control for a silicon microactuator, Sensors and Actuator A21–A23 (1990), pp. 229–235

[30] M.W. Hamberg, C. Neagu, J.G.E. Gardeniers, D.J. Ijntema, and M. Elwenspoek, An electrochemical microactuator, Proc. MEMS 95, Amsterdam, the Netherlands, IEEE Catalog No. 95CH35754 (1995), pp. 106–110

[31] N.E. Agbor, M.C. Petty, and A.P. Monkman, Polyaniline thin films for gas sensing, Sensors and Actuators B28 (1995), pp. 173–179

[32] S. Shoji and M. Esashi, Microflow devices and systems, J. Micromech. Microeng. 4 (1994), pp. 157–171

[33] H. Jerman, Electrically-activated micromachined diaphragm valves, Proc. Microsystem Technologies 90, Berlin, Germany (1990), pp. 806–811

[34] H. Jerman, Electrically-activated, normally-closed diaphragm valves, Proc. Transducers 91, San Francisco, USA (1991), pp. 1045–1048

[35] H.P. Trah, H. Baumann, C. Döring, H. Goebel, T. Grauer, and M. Mettner, Micromachined valve with hydraulically actuated membrane subsequent to a thermoelectrically controlled bimorph cantilever, Sensors and Actuators A 39 (1993), pp. 169–176

[36] H. Jerman, Electrically-activated, normally-closed diaphragm valve, J. Micromech. Microeng. 4 (1994), pp. 210–216

[37] M. Freygang, H. Haffner, S. Messner, and B. Schmidt, A new concept of a bimetallically actuated, normally-closed microvalve, Proc. Transducers 95, Stockholm, Sweden (1995), p. 530-News

[38] M. Freygang, H. Glosch, H. Haffner, and S. Messner, Thermal actuation units for microvalves and micropumps, Proc. Actuator 96, Bremen, Germany (1996), pp. 84–87

[39] J. Franz, H. Baumann, and H.P. Trah, A monolithic silicon micro-liquid handling system with bimetallic valve and calorimetric flow sensor, Proc. Actuator 96, Bremen, Germany (1996), pp. 29–32

[40] S. Messner, M. Müller, V. Burger, J. Schaible, and R. Zengerle, A 3-way microvalve for pneumatic applications, Proc. Actuator 98, Bremen, Germany (1998), pp. 107–110

[41] M.J. Zdeblick and J.B. Angell, A microminiature electric-to-fluidic valve, Proc. Transducers 87, Tokyo, Japan (1987), pp. 827–829

[42] M.J. Zdeblick, R. Anderson, J. Jankowski, B. Kline-Schoder, L. Christel, R. Miles, W. Weber, Thermopneumatically actuated microvalve and integrated electro-fluidic circuits, Proc. Actuator 94, Bremen, Germany (1994), pp. 56–60

[43] J. Fahrenberg, D. Maas, W. Menz, and W.K. Schomburg, Active microvalve system manufactured by the LIGA process, Proc. Actuator 94, Bremen, Germany (1994), pp. 71–74

[44] C.A. Rich and K.D. Wise, An 8-bit microflow controller using pneumatically-actuated microvalves, Proc. MEMS 99, Orlando, Florida, USA (1999), pp. 130–135

[45] C. Grosjean, X. Yang, and Y.C. Tai, A practical thermopneumatic valve, Proc. MEMS 99, Orlando, Florida, USA (1999), pp. 147–152

[46] T. Lisec, S. Hoerschelmann, H.J. Quenzer, B. Wagner, and W. Benecke, Thermally driven microvalve with buckling behavior for pneumatic applications, Proc. MEMS 94, Oiso, Japan, IEEE Catalog No. 94CH3404-1, (1994), pp. 13–17

[47] C. Goll, W. Bacher, B. Büstgens, D. Maas, W. Menz, and W.K. Schomburg, Microvalves with bistable buckled polymer diaphragms, Proc. Transducers 95, Stockholm, Sweden (1995), p. 523-News
[48] C. Goll, W. Bacher, B. Büstgens, D. Maas, W. Menz, and W.K. Schomburg, Microvalves with bistable buckled polymer diaphragms, J. Micromech. Microeng. 6 (1996), pp. 77–79
[49] T. Lisec, M. Kreutzer, and B. Wagner, A bistable microswitch for driving fluidic components, Sensors and Actuators A 54 (1996), pp. 746–749
[50] C. Goll, Entwicklung, Herstellung und Test von aktiven Mikroventilen für pneumatische Anwendungen, thesis, University of Karlsruhe (1997)
[51] M.J. Tsai, S.C. Hsu, C.H. Ke, R.H. Jang, and C.Y. Wu, A thermally buckling microvalve with integrated flow sensor, Proc. Actuator 98, Bremen, Germany (1998), pp. 130–136
[52] M. Esashi, Integrated microflow control systems, Sensors and Actuators A21–A23 (1990), pp. 161–167
[53] S. Shoji, B. van der Shoot, N.F. de Rooij, and M. Esashi, Smallest dead volume microvalves for integrated chemical analyzing systems, Proc. Transducers 91, San Francisco, USA (1991), pp. 1052–1055
[54] J. Joswig, Active micromechanic valve, J. Micromech. Microeng. 2 (1992), pp. 262–265
[55] J. Shinohara, M. Suda, K. Furuta, and T. Sakuhara, A high pressure-resistance micropump using active and normally-closed valves, Proc. MEMS 00, Miyazaki, Japan (2000)
[56] M. Weinmann, P. Post, H. Vollmer, R. Wanner, S. Kluge, and P. Woias, Pneumatic silicon microvalves with piezoelectric actuation, Proc. Actuator 00, Bremen, Germany (2000), pp. 224–227
[57] T. Ohnstein, T. Fukiura, J. Ridley, and U. Bonne, Micromachined silicon microvalve, Proc. MEMS 90, Napa Valley, USA (1990), pp. 95–98
[58] J. Branebjerg and P. Gravesen, A new electrostatic actuator providing improved stroke-length and force, Proc. MEMS 92, Travemünde, Germany, IEEE Catalog No. 92CH3093-2, (1992), pp. 6–11
[59] D. Bosch, B. Heimhofer, G. Muck, H. Seidel, U. Thumser, and W. Welser, A silicon microvalve with combined electromagnetic/electrostatic actuation, Sensors and Actuators A37–A38 (1992), pp. 684–692
[60] M.A. Huff, J.R. Gilbert, and M.A. Schmidt, Flow characteristics of a pressure-balanced microvalve, Proc. Transducers 93, Yokohama, Japan (1993), pp. 98–101
[61] M. Shikida and K. Sato, Characteristics of an electrostatically driven gas valve under high pressure conditions, Proc. MEMS 94, Oiso, Japan, IEEE Catalog No. 94CH3404-1, (1994), pp. 235–240
[62] K. Sato and M. Shikida, An electrostatically actuated gas valve with an S-shaped film element, J. Micromech. Microeng. 4 (1994), pp. 205–209

[63] J.H. Babaei, R.S. Huang, and Ch.Y. Kwok, A new bistable microvalve using a SiO_2 beam as the movable part, Proc. Actuator 94, Bremen, Germany (1994), pp. 34–37

[64] C. Goll, W. Bacher, B. Büstgens, D. Maas, R. Ruprecht, and W.K. Schomburg, Electrostatically actuated polymer microvalve equipped with a movable membrane electrode, Proc. MME 96, Barcelona, Spain (1996), pp. 231–234

[65] S. Kluge, G. Klink, M. Biller, B. Hillerich, and P. Woias, On the static and dynamic behaviour of an electrostatically movable plate actuator, Proc. Actuator 98, Bremen, Germany (1998), pp. 114–117

[66] K. Yanagisawa, H. Kuwano, and A. Tago, An electromagnetically driven microvalve, Proc. Transducers 93, Yokohama, Japan (1993), pp. 102–105

[67] O. Krusemark, A. Feustel, and J. Müller, Electromagnetic active valve for fluids and gases, Proc. Actuator 98, Bremen, Germany (1998), pp. 126–128

[68] A. Meckes, J. Behrens, and W. Benecke, A microvalve with electromagnetic actuator, Proc. Actuator 98, Bremen, Germany (1998), pp. 152–155

[69] S. Shoji, M. Esashi, and M. Matsuo, Prototype miniature blood gas analyzer fabricated on a silicon wafer, Sensors and Actuators 14 (1988), pp. 101–107

[70] J.D. Bush and A.D. Johnson, Prototype microvalve actuator, Proc. MEMS 90, Napa Valley, California, USA (1990)

[71] C.A. Ray, C.L. Sloan, A.D. Johnson, J.D. Busch, and B.R. Petty, A silicon-based shape memory alloy microvalve, Mat. Res. Soc. Symp. Proc. Vol. 276 (1992), pp. 161–166

[72] A.D. Johnson and A.R. Curtis, Shape memory alloy film actuated microvalve, U.S. Patent #5,325,880, July 5, (1994)

[73] A.D. Johnson and E.J. Shahoian, Recent progress in thin film shape memory microactuators, Proc. MEMS 95, Amsterdam, the Netherlands, IEEE Catalog No. 95CH35754, (1995), pp. 216–220

[74] C.R. Neagu, J.G.E. Gardeniers, M. Elwenspoek, and J.H.J. Fluitman, An electrochemically actuated microvalve, Proc. Actuator 96, Bremen, Germany (1996), pp. 41–44

[75] T.S.J. Lammerink, N. Olij, J.W. Berenschot, R.E. Oosterbroek, J.G.E. Gardeniers, C. Neagu, A. van den Berg, M. Haller, G.J.M. Krijnen, and M.C. Elwenspoek, A normally-closed active valve using bulk micromachining, Proc. Actuator 98, Bremen, Germany (1998), pp. 122–125

[76] Voltage sensitive orifice, LDI Pneutronics, Hollis, NH, USA

[77] H. Töpfer, Bürkert Inc., Großröhrsdorf, Germany, private communication

[78] P. Muralt, M. Kohli, T. Maeder, A. Kholkin, K. Brooks, N. Setter, and R. Luthier, Fabrication and characterization of PZT thin-film vibrators for micromotors, Sensors and Actuators A 48 (1995), pp. 157–165

[79] M. Mescher, T. Abe, B. Brunett, H. Metla, T.E. Schlesinger, and M. Reed, Piezoelectric lead-zirconate-titanate actuator films for microelectrome-

chanical systems applications, Proc. MEMS 95, Amsterdam, the Netherlands, IEEE Catalog No. 95CH35754, (1995), pp. 261–266

[80] F. Gugenberger, I. Fergen, E. Quandt, and H. Holleck, Materials development of piezoelectric thin films for actuator devices – comparative study of PZT and wurzite type materials grown by magnetron sputtering, Proc. Micro Materials 97, Berlin, Germany (1997), pp. 608–611

[81] Fluistor^{TM}, Redwood Microsystems, Menlo Park, CA, USA

[82] EG&G IC Sensors, Milpitas, CA, USA

[83] B. Wagner, M. Kreutzer, and W. Benecke, Linear and rotational magnetic micromotors fabricated using silicon technology, Proc. MEMS 92, Travemünde, Germany, IEEE Catalog No. 92CH3093-2, (1992), pp. 183–189

[84] H. Guckel, T. Christensen, T. Earles, K. Skrobis, D. Zook, and T. Ohnstein, Electromagnetic, spring constrained linear actuator with large throw, Proc. Actuator 94, Bremen, Germany (1994), pp. 52–55

[85] H. Guckel, T. Earles, J. Klein, D. Zook, and T. Ohnstein, Electromagnetic linear actuators with inductive position sensing for microrelay, microvalve and precision positioning applications, Proc. Transducers 95, Stockholm, Sweden (1995), Vol. 1, pp. 324–327

[86] B. Rogge, J. Schulz, J. Mohr, A. Thommes, and W. Menz, Fully batch fabricated magnetic microactuators using a two layer LIGA process, Proc. Transducers 95, Stockholm, Sweden (1995), Vol. 1, pp. 320–323

[87] B. Rogge, J. Schulz, J. Mohr, and A. Thommes, Magnetic microactuators fabricated by the LIGA technique for large displacements or large forces, Proc. Actuator 96, Bremen, Germany (1996), pp. 112–115

[88] S. Massoud-Ansari, P.S. Magnat, J. Klein, and H. Guckel, A multi-level, LIGA-like process for three dimensional actuators, Proc. MEMS 96, San Diego, USA (1996), pp. 285–289

[89] T. Earles, P. Mangat, J. Klein, and H. Guckel, Magnetic microactuators for relay applications, Proc. Actuator 96, Bremen, Germany (1996), pp. 132–135

[90] H. Guckel, Progress in electromagnetic microactuators, Proc. Actuator 96, Bremen, Germany (1996), pp. 45–48

[91] N. Tabat, J. Klein and H. Guckel, Single flux-path bidirectional linear actuators, Proc. Transducers 97, Chicago, USA (1997), pp. 789–792

[92] M. Klöpzig, A novel linear micromachined electromagnetic actuator including magnetic suspension, Proc. Actuator 98, Bremen, Germany (1998), pp. 548–551

[93] P. Krippner, Hochempfindliche LIGA-Mikrospektrometersysteme für den Infrarotbereich, thesis, University of Karlsruhe (1999)

[94] W.C. Tang, T.H. Nguyen, and R.T. Howe, Laterally driven polysilicon resonant microstructures, Sensors and Actuators 20 (1989), pp. 25–32

[95] J. Mohr, C. Burbaum, P. Bley, W. Menz, and U. Wallrabe, Movable microstructures manufactured by the LIGA process as basic elements for micro-

systems, Proc. Int. Conf. on Microelectromechanical Systems, Berlin, Germany (1990), pp. 529–538

[96] C.J. Kim, A.P. Pisano, R.S. Muller, and M.G. Lim, Polysilicon microgripper, IEEE Solid-State Sensor and Actuator Workshop, IEEE Catalog No. 90CH2783-9, (1990), pp. 48–51

[97] T. Hirano, T. Furuhata, K.J. Gabriel, and H. Fujita, Design, fabrication and operation of submicrometer gap comb-drive microactuators, J. Microelectromech. 1 (1992), pp. 52–59

[98] D. Kobayashi, T. Hirano, T. Furuhata, and H. Fujita, An integrated lateral tunneling unit, Proc. MEMS 92, Travemünde, Germany, IEEE Catalog No. 92CH3093-2, (1992), pp. 214–219

[99] Y. Gianchandani and K. Najafi, Micron-sized, high aspect ratio bulk silicon micromechanical devices, Proc. MEMS 92, Travemünde, Germany, IEEE Catalog No. 92CH3093-2, (1992), pp. 208–213

[100] V.P. Jaecklin, C. Linder, N.F. de Rooij, and J.M. Moret, Micromechanical comb actuators with low driving voltage, J. Micromech. Microeng. 2 (1992), pp. 250–255

[101] V.P. Jaecklin, C. Linder, N.F. de Rooij, J.M. Moret, R. Bischof, and F. Rudolf, Novel polysilicon comb-actuators for XY-stages, Proc. MEMS 92, Travemünde, Germany, IEEE Catalog No. 92CH3093-2, (1992), pp. 147–149

[102] J. Brugger, V.P. Jaecklin, C. Linder, N. Blanc, P.F. Indermühle, and N.F. de Rooij, Microfabricated tools for nanoscience, J. Micromech. Microeng. 3 (1993), pp. 161–167

[103] M. Kohl and J. Mohr, Entwurf elektrostatischer Linearaktoren, IMT report No. 110/30 (1993)

[104] J. Mohr, M. Kohl, and W. Menz, Micro optical switching by electrostatic linear actuators with large displacements, Proc. Transducers 93, Yokohama, Japan (1993), pp. 120–124

[105] T. Akiyama and K. Shono, Controlled stepwise motion in polysilicon microstructures, J. Microelectromechanical Syst., Vol. 2, No. 3 (1993), pp. 106–110

[106] H. Kalb, B. Kowanz, W. Bacher, J. Mohr, and R. Ruprecht, Electrostatically driven linear stepping motor in LIGA technique, Proc. Actuator 94, Bremen, Germany (1994), pp. 83–85

[107] H. Guckel, T.R. Christenson, T. Earles, K.J. Skrobis, and J. Klein, Processing and design considerations for high force output – large throw electrostatic, linear microactuators, Proc. Actuator 94, Bremen, Germany (1994), pp. 105–108

[108] A. Selvakumar, K. Najafi, W.H. Juan, and S. Pang, Vertical comb array microactuators, Proc. MEMS 95, Amsterdam, the Netherlands, IEEE Catalog No. 95CH35754, (1995), pp. 43–48

[109] M.W. Börner, M. Kohl, F.J. Pantenburg, W. Bacher, H. Hein, and W.K. Schomburg, Movable microstructures made by a sub-micron LIGA process, Proc. HARMST 95, Karlsruhe, Germany (1995)
[110] M.W. Börner, M. Kohl, F.J. Pantenburg, W. Bacher, H. Hein, and W.K. Schomburg, Sub-micron LIGA process for movable microstructures, Microelectronic Engineering 30 (1996), pp. 505–508
[111] M. Kohl, J. Göttert, and J. Mohr, Verification of the micromechanical characteristics of electrostatic linear actuators, Sensors and Actuators A 53 (1996), pp. 416–422
[112] U. Gengenbach, J. Göttert, M. Kohl, M. Marth, J. Mohr, and A. Müller, A modular fabrication concept for photonic devices using the LIGA process, Proc. Microsystem Technologies 96, Potsdam, Germany (1996), pp. 519–524
[113] M.H. Kiang, J.T. Nee, K.Y. Lau, and R.S. Muller, Surface-micromachined diffraction gratings for scanning spectroscopic applications, Proc. Transducers 97, Chicago, USA (1997), pp. 343–345
[114] N. Tas, J. Wissink, L. Sander, T. Lammerink, and M. Elwenspoek, The shuffle motor: a high force, high precision linear electrostatic stepper motor, Proc. Transducers 97, Chicago, USA (1997), pp. 777–780
[115] M. Baltzer, T. Kraus, and E. Obermeier, A linear stepping actuator in surface micromachining technology for low voltages and large displacements, Proc. Transducers 97, Chicago, USA (1997), pp. 781–784
[116] H. Goto, Y. Kamoda, and M. Yoneda, Miniature two-dimensional optical device utilizing a multilayered piezoelectric actuator with new displacement expansion mechanism, Proc. Actuator 92, Bremen, Germany (1992), pp. 134–140
[117] A.M. Flynn, L.S. Tavrow, S.F. Bart, R.A. Brooks, D.J. Ehrlich, K.R. Udayakumar, and L.E. Cross, Piezoelectric micromotors and microrobots, J. Microelectromechanical Systems, Vol.1, No. 1 (1992), pp. 44–51
[118] K. Spanner and W.W. Wolny, Trends and challenges in new piezoelectric actuator applications, Proc. Actuator 96, Bremen, Germany (1996), pp. 140–146
[119] P.D. Atherton, and K. Uchino, New developments in piezo motors and mechanisms, Proc. Actuator 98, Bremen, Germany (1998), pp. 164–169
[120] H. Debeda, J. Mohr, C. Rembe, J. Schulz, U. Wallrabe, and J. Wengelink, Fully batch fabricated LIGA actuators integrated on piezoelectric ceramic substrates, Proc. Actuator 98, Bremen, Germany (1998), pp. 54–57
[121] M. Hermann, W. Schinköthe, and J. Haug, Properties of a piezoelectric travelling wave motor generating direct linear motion, Proc. Actuator 98, Bremen, Germany (1998), pp. 238–241
[122] J.M. Breguet and R. Clavel, New designs for long range, high resolution, multi-degrees-of-freedom piezoelectric actuators, Proc. Actuator 98, Bremen, Germany (1998), pp. 198–201

[123] E. Quandt, A. Ludwig, and K. Seemann, Giant magnetostrictive multilayers for thin film actuators, Proc. Transducers 97, Chicago, USA (1997), pp. 1089–1092
[124] F. Claeyssen, N. Lhermet, J. Betz, K. MacKay, D. Givord, E. Quandt, and H. Kronmüller, Linear and rotating magnetostrictive micromotors, Proc. Actuator 98, Bremen, Germany (1998), pp. 372–375
[125] J.S. Park, L.L. Chu, E. Siwapornsathain, A.D. Oliver, and Y.B. Gianchandani, Long throw and rotary output electro-thermal actuators based on bent-beam suspensions, Proc. MEMS 00, Miyazaki, Japan (2000)
[126] X. Yang, Y.C. Tai, and C.M. Ho, Micro bellows actuators, Proc. Transducers 97, Chicago, USA (1997), pp. 45–48
[127] T.R. Christenson, T.J. Garino, E.L. Venturini, and D.M. Berry, Application of deep X-ray lithography fabricated rare-earth permanent magnets to multipole magnetic microactuators, Proc. Transducers 99, Sendai, Japan (1999), pp. 98–101
[128] T.M. Liakopoulos, W. Zhang, and C.H. Ahn, Electroplated thick CoNiMnP permanent magnet arrays for micromachined magnetic device applications, Proc. MEMS 96, San Diego, USA (1996), pp. 79–84
[129] A. Thommes, W. Bacher, K. Leyendecker, W. Stark, H. Liebscher, and Ch. Jakob, LIGA-microstructures from NiFe alloy: preparation by electroforming and their magnetic properties, Proc. Meeting of the American Electrochem Soc., New Orleans, USA (1993), pp. 89–102
[130] W.S.N. Trimmer and K.J. Gabriel, Design considerations for a practical electrostatic micromotor, Sensors and Actuators 11 (1987), pp. 189–206
[131] T.A. Roessig, R.T. Howe, A.P. Pisano, and J.H. Smith, Surface-micromachined resonant accelerometer, Proc. Transducers 97, Chicago, USA (1997), p. 3B2.02
[132] J.H.J. Fluitman and H. Guckel, Micro actuator principles, Proc. Actuator 96, Bremen, Germany (1996), pp. 23–28
[133] K. Spanner, Breakthrough in piezo actuator applications, Proc. Actuator 00, Bremen, Germany (2000), pp. 236–241
[134] C.M. Wayman, Introduction to Crystallography of Martensitic Transformations, MacMillan, New York (1964), p. 23
[135] J.W. Christian, The Theory of Transformations in Metals and Alloys, Pergamon Press, Oxford (1965)
[136] Z. Nishiyama, Martensitic Transformation, Academic Press, New York (1978)
[137] E. Hornbogen, in: 'Legierungen mit Formgedächtnis', Expert, Esslingen, Ed. D. Stöckel (1988)
[138] W.J. Buehler, J.V. Gilfrich, and R.C. Wiley, J. Appl. Phys. 34 (1963) p. 1475
[139] K. Otsuka and C.M. Wayman, Shape Memory Materials, Cambridge University Press (1998)

[140] S. Miyazaki and K. Otsuka, Deformation and Transition Behavior Associated with the R-Phase in Ti-Ni Alloys, Metall. Trans. A, Vol. 17A, (1986), pp. 53–63
[141] K. Takezawa, H. Imamura, K. Tanizaki, and S. Sato, J. de Physique C-4 Suppl. No. 12 (1982), pp. 741–746
[142] S. Miyazaki, S. Kimura, K. Otsuka, and Y. Suzuki, Scr. Metall. 18 (1984), p. 883
[143] R. Kainuma, M. Matsumoto, and T. Honma, Proc. ICOMAT '86, Nara, Japan (1986), p. 717
[144] T. Saburi, S. Nenno, Y. Nishimoto, and M. Zeniya, J. Iron and Steel Inst. Japan, 72 (1986), p. 571
[145] K. Otsuka, H. Sakamoto, and K. Shimizu, Acta Metall. 27 (1979), p. 585
[146] T.W. Duerig and R. Zadno, An engineer's perspective of pseudoelasticity, in: Engineering Aspects of Shape Memory Alloys, Butterworth-Heinemann, London (1990), pp. 369–393
[147] E. Hornbogen, On the term pseudoelasticity, Z. Metall. 86 (1995), pp. 341–344
[148] I. Müller and H. Xu, On the hysteresis in shape memory alloys, in: The Martensitic Transformation in Science and Technology, Eds. E. Hornbogen and N. Jost, DGM Informationsgesellschaft Verlag (1989), pp. 319–324
[149] R.W. Cahn, Acta Metall., 1 (1953), p. 49
[150] K. Otsuka and C.M. Wayman, in: Deformation Behavior of Materials, Ed. Feltham, Vol. II, Freund Publishing House, Israel (1977), p. 91
[151] S. Miyazaki, S. Kimura, and K. Otsuka, Scripta Metall. 18 (1984), p. 883
[152] T. Saburi and S. Nenno, Proc. Int. Conf. on Solid-Solid Phase Transformations, AIME, New York, (1982), p. 1455
[153] G. Guenin, The two-way memory effect in relation to the nucleation and growth of martensite, in: The Martensitic Transformation in Science and Technology, Eds. E. Hornbogen and N. Jost, DGM Informationsgesellschaft Verlag (1989), pp. 39–52
[154] S. Miyazaki, Y. Ohmi, K. Otsuka, and Y. Suzuki, J. de Physique 43, Suppl. 12 (1982), p. C4-255
[155] M.S. Wechsler, D.S. Lieberman, and T.A. Read, Trans. AIME, 197 (1953), p. 1503
[156] J.S. Bowles and J.K. Mackenzie, Acta Metall., 2 (1954), p. 129
[157] J.W. Christian, J. Inst. Metals, 84 (1955), p. 386
[158] J. Tichy and G. Gautschi, Piezoelektrische Meßtechnik, Kap. 4, Elastische Eigenschaften der Kristalle, Springer, Heidelberg (1980)
[159] J. Ortin and A. Planes, Acta Metall. 36 (1988), p. 1873
[160] E. Hornbogen, The Effect of Variables on Martensitic Transformation Temperatures, Acta. Metall. 33 (1985), pp. 595–601
[161] X. Gao, M. Huang, and L.C. Brinson, A multivariant micromechanical model for SMAs. Part 1. Crystallographic issues for single crystal model, Int. J. of Plasticity 16 (2000), pp. 1345–1369

[162] M. Huang, X. Gao, and L.C. Brinson, A multivariant micromechanical model for SMAs. Part 2. Polycrystal model, Int. J. of Plasticity 16 (2000), pp. 1371–1390

[163] M. Achenbach and I. Müller, Simulation of material behaviour of alloys with shape memory, Arch. Mech. 37, 6 (1985), pp. 573–585

[164] N. Papenfuß and S. Seelecke, Simulation and control of SMA actuators, Proc. SPIE 6th Ann. Int. Symp. Smart Struct. Mat., Vol. 3667, Newport Beach, USA (1999)

[165] K. Tanaka, S. Kobayashi, and Y. Sato, Thermomechanics of transformation pseudoelasticity and shape memory effect in alloys, Int. J. Plasticity, 2 (1986), pp. 59–72

[166] C. Liang and C.A. Rogers, One-dimensional thermomechanical constitutive relations for shape memory materials, J. Intell. Matl. Syst. & Struct., 1 (1990), pp. 207–234

[167] L.C. Brinson, One-dimensional constitutive behavior of shape memory alloys: thermomechanical derivation with non-constant material functions and redefined martensite internal variable, J. Intell. Matl. Syst. & Struct., 4 (1993), pp. 229–242

[168] K. Ikuta and H. Shimizu, Two-dimensional mathematical model of shape memory alloy and intelligent SMA-CAD, Proc. MEMS 93, Fort Lauderdale, USA, IEEE Catalog No. 0-7803-0957-2/93 (1993), pp. 87–92

[169] J.G. Boyd and D.C. Lagoudas, Thermomechanical response of shape memory composites, J. Intell. Matl. Syst. & Struct., 5 (1994), pp. 333–346

[170] Y. Ivshin and T.J. Pence, A thermomechanical model for a one variant shape memory material, J. Intell. Matl. Syst. & Struct., 5 (1994), pp. 455–473

[171] L.C. Brinson and M.S. Huang, Simplifications and comparisons of shape memory alloy constitutive models, J. Intell. Matl. Syst. & Struct., 7 (1996), pp. 108–114

[172] S. Leclercq and C. Lexcellent, A general macroscopic description of the thermomechanical behavior of shape memory alloys, J. Mech. Phys. Solids, Vol. 44, No.6 (1996), pp. 953–980

[173] P. Wollants, M. De Bonte, and J.R. Roos, Z. Metallkd., 70 (1979), p. 113

[174] G.B. Stachowiak and P.G. McCormic, Shape memory behaviour associated with the R and martensitic transformations in a NiTi alloy, Acta Metall., Vol. 36 No. 2 (1988), pp. 291–297

[175] B. Krevet and M. Kohl, Simulation of shape memory devices with coupled finite element programs, Proc. of MSM 2001, Florida, USA (2001), pp. 282–285

[176] L. Juhasz, H. Andrä, and E. Schnack, Modeling of the general nonproportional loading and the reorientation in a polycrystalline SMA, Proc. workshop on 'Shape memory alloys – experimental verification and numerical modeling', Karlsruhe, Germany (2001)

[177] C. Bouvet, S. Calloch, and Ch. Lexcellent, Experimental and theoretical investigations of shape memory polycrystalline alloys: thermomechanical behavior under multiaxial loadings, NATO advanced research workshop, Metz, France (2002), to be published in NATO ASI Series by Kluver Academic Press

[178] K.N. Melton, Ni-Ti based shape memory alloys, in: Engineering Aspects of Shape Memory Alloys, Butterworth-Heinemann, London (1990), pp. 369–393

[179] G. Chattopadhyay and H. Kleykamp, Z. Metallkd. 74 (1983), p. 182

[180] T. Saburi, M. Yoshida, and S. Nenno, Scr. Metall. 18 (1984), p. 363

[181] G.D. Sandrock, A.J. Perkins, and R.F. Hehemann, Metall. Trans. 2 (1971), p. 2769

[182] M. Kohl, K.D. Skrobanek, A. Schüßler, M. Teschner, and S. Miyazaki, Stress-optimised integrated linear actuators with shape memory effect, Proc. Actuator 96, Bremen, Germany (1996), pp. 366–369

[183] M. Kohl, E. Just, W. Pfleging, and S. Miyazaki, SMA microgripper with integrated antagonism, Sensors and Actuators A83 (2000), pp. 208–213

[184] M. Kohl, D.M. Allen, T.T. Chen, S. Miyazaki, and M. Schwörer, Anisotropy in microdevices produced by micromachining of cold-rolled NiTi sheets, Materials Science and Engineering A270 (1999), pp. 145–150

[185] M. Kohl, K.D. Skrobanek, C.M. Goh, and D.M. Allen, Mechanical characterization of shape memory micromaterials, Proc. Int. Conf. on Microlithography and Metrology in Micromachining II, Austin, Texas, USA, Eds. M.T. Postek and C. Friedrich, SPIE Vol. 2880, (1996), pp. 108-118

[186] K.D. Skrobanek, M. Kohl, and S. Miyazaki, Stress-optimised shape memory devices for the use in microvalves, Proc. ESOMAT 97, Enschede, the Netherlands (1997); J. Phys. IV France 7 (1997), Colloque C5, pp. 597–602

[187] K. Kitamura, S. Miyazaki, H. Iwai, and M. Kohl, Effect of heat treatment on the texture in rolled TiNi thin plates, Proc. SMST 97, Pacific Grove, California, USA (1997), p. 47

[188] A. Khantachawana, S. Miyazaki, H. Iwai, and M. Kohl, Effect of heat treatment and anisotropy of transformation strain in Ti-Ni-Fe rolled thin plates, Mat. Science and Eng., A273–275 (1999), pp. 763–768

[189] K. Kitamura, S. Miyazaki, H. Iwai, and M. Kohl, Effect of rolling reduction on the deformation texture and anisotropy of transformation strain in Ti-50.2 at.% Ni thin plates, Mat. Science and Eng., A273–275 (1999), pp. 758–762

[190] D. Stöckel, Status and trends in shape memory technology, Proc. Actuator 92, Bremen, Germany (1992) pp. 79–84

[191] S. Miyazaki, Thermal and stress cycling effects and fatigue properties of NiTi alloys, in: Engineering Aspects of Shape Memory Alloys, Butterworth-Heinemann, London (1990), pp. 394–413

[192] W.J. Moberly and K.N. Melton, Ni-Ti-Cu shape memory alloys, in: Engineering Aspects of Shape Memory Alloys, Butterworth-Heinemann, London (1990), pp. 46–57
[193] M. Kohl and K.D. Skrobanek, Linear microactuators based on the shape memory effect, Proc. Transducers 97, Chicago, USA, IEEE Catalog No. 97TH8267, (1997), pp. 785–788
[194] P.G. Lindquist and C.M. Wayman, Shape memory and transformation behavior of martensitic Ti-Pd-Ni and Ti-Pt-Ni alloys, in: Engineering Aspects of Shape Memory Alloys, Butterworth-Heinemann, London (1990), pp. 58–68
[195] V.N. Khachin, V.E. Grunter, V.P. Sivokha, and A.S. Savvinov, Proc. ICOMAT 79, Cambridge, MA, USA (1979), p. 474
[196] K. Ikuta, M. Hayashi, T. Matsuura, and H. Fujishiro, Shape memory alloy thin film fabricated by laser ablation, Proc. MEMS 94, Oiso, Japan, IEEE Catalog No. 94CH3404-1, (1994), pp. 355–360
[197] S. Kajiwara, T. Kikuchi, and K. Ogawa, Strengthening of Ti-Ni shape memory films by coherent subnanometric plate precipitates, Phil. Mag. Letters, Vol. 74, No. 3, (1996), pp. 137–144
[198] A. Kajiwara, K. Yamazaki, K. Ogawa, T. Kikuchi, and S. Miyazaki, Basic principles for alloy designing of composition-insensitive SMA Ti-Ni based thin films with large recovery force, Proc. SMART 2000, Sendai, Japan
[199] M. Kohl, D. Dittmann, E. Quandt, and B. Winzek, Thin film shape memory microvalves with adjustable operation temperature, Sensors and Actuators A83 (2000), pp. 214–219
[200] M. Kohl, K.D. Skrobanek, E. Quandt, P. Schloßmacher, A. Schüßler, and D.M. Allen, Development of microactuators based on the shape memory effect, Journal de Physique IV C8 (1995), pp. 1187–1192
[201] J. Hesselbach, R. Pittschellis, H. Stork, E. Hornbogen, and M. Mertmann, Optimization and control of electrically heated shape memory actuators, Proc. Actuator 94, Bremen, Germany, Ed. H. Borgmann, (1994), pp. 337–340
[202] T. Hashinaga, S. Miyazaki, T. Ueki, and H. Horikawa, Transformation and deformation behavior in sputter-deposited Ti-Ni-Cu thin films, Journal de Physique IV C8 (1995), pp. 689–694
[203] E. Quandt and H. Holleck, Shape memory thin films of the system Ti-(Ni-Pd-Cu), Mat. Res. Soc. Symp. Proc., Vol. 459 (1997), pp. 465–470
[204] E. Quandt, C. Halene, H. Holleck, K. Feit, M. Kohl, P. Schloßmacher, A. Skokan, and K.D. Skrobanek, Sputter deposition of TiNi, TiNiPd and TiPd films displaying the two-way shape memory effect, Sensors and Actuators A 53 (1996), pp. 434–439
[205] S. Miyazaki, T. Hashinaga, K. Yumikura, H. Horikawa, T. Ueki, and A. Ishida, Shape memory characteristics of sputter-deposited Ti-Ni-base thin films, SPIE Proc. Series Vol. 2441 (1995), pp. 156–164

[206] M Kohl, D. Dittmann, E. Quandt, B. Winzek, S. Miyazaki, and D.M. Allen, Shape memory microvalves based on thin films or rolled sheets, Mat. Science & Eng. A273–275 (1999), pp. 784–788
[207] B. Winzek, E. Quandt, and H. Holleck, Shape memory thin film composites with adjustable transformation temperatures and hysteresis, Proc. Actuator 98, Bremen, Germany, Ed. H. Borgmann, (1998), pp. 461–464
[208] B. Winzek, Entwicklung, Herstellung und Charakterisierung von Mikroaktoren mit Formgedächtnisschichten auf der Basis von NiTi, thesis, University of Karlsruhe (2000)
[209] B. Winzek, E. Quandt, and H. Holleck, Thin Ti(Ni,Cu) film composites, Proc. Actuator 00, Bremen, Germany, Ed. H. Borgmann, (2000), pp. 172–176
[210] K. Ullakko, J.K. Huang, C. Kantner, R.C. O'Handley, and V.V. Kokorin, Appl. Phys. Lett., Vol. 69 (1996), pp. 1966–1968
[211] D.W. Pepper and J.C. Heinrich, The Finite Element Method, Hemisphere Publishing Corporation, Washington (1992)
[212] S.S. Rao, The Finite Element Method in Engineering, Pergamon Press, Oxford (1989)
[213] L.C. Brinson and R. Lammering, Finite element analysis of the behavior of shape memory alloys and their applications, Int. J. Solids Structures, Vol. 30, No. 23, (1993), pp. 3261–3280
[214] F. Auricchio and R.L. Taylor, Shape memory alloy superelastic behavior: 3D finite element simulations, Proc. ICIM '96, Lyon, France, SPIE Vol. 2779, (1996), pp. 487–492
[215] S. Leclercq, C. Lexcellent, and J.C. Gelin, A finite element calculation for the design of devices made of shape memory alloys, Journal de Physique IV, Colloque C1, Vol. 6, (1996), pp. 225–234
[216] S. Seelecke and N. Papenfuß, A finite element formulation for SMA actuators, Proc. 4th Europ. Conf. Smart Struct. Mat., Harrogate, UK, (1998), pp. 357–363
[217] B. Krevet and M. Kohl, Gekoppelte Simulation für Anwendungen in der Mikrosystemtechnik, Proc. of 4th PMT-Statuskolloquium, Forschungszentrum Karlsruhe (2000), p. 209
[218] B. Krevet and W. Kaboth, Coupling of FEM programs for simulation of complex systems, Proc. MSM 98, Santa Clara, USA (1998), pp. 320–32
[219] O.C. Zienkiewicz, The Finite Element Method, McGraw-Hill Book Company, 3rd Ed. (1977), pp. 500–523
[220] K.D. Skrobanek, O.F. Hagena, and M. Kohl, Entwicklung von Mikromembranaktoren mit NiTi-Formgedächtnislegierungen, thesis, University of Karlsruhe (1998)
[221] C. Mattheck and S. Burkhardt, Int. J. Fatigue, Vol. 12, (1990), pp. 185–190
[222] M. Kapgan and K.N. Melton, Shape memory alloy tube and pipe couplings, in: Engineering Aspects of Shape Memory Alloys, Eds. T.W. Duerig, K.N.

Melton, D. Stöckel, and C.M. Wayman, Butterworth-Heinemann, London (1990), pp. 137–148

[223] E. Cydzik, The design of electrical interconnection systems with shape memory alloys, in: Engineering Aspects of Shape Memory Alloys, Eds. T.W. Duerig, K.N. Melton, D. Stöckel, and C.M. Wayman, Butterworth-Heinemann, London (1990), pp. 149–157

[224] S.M. Fisher, In line anti-scald safety valve, in: Engineering Aspects of Shape Memory Alloys, Eds. T.W. Duerig, K.N. Melton, D. Stöckel, and C.M. Wayman, Butterworth-Heinemann, London (1990), pp. 303–314

[225] Y. Furuya and H. Shimada, Shape memory actuators for robotic applications, in: Engineering Aspects of Shape Memory Alloys, Eds. T.W. Duerig, K.N. Melton, D. Stöckel, and C.M. Wayman, Butterworth-Heinemann, London (1990), pp. 338–355

[226] T. Borden, Shape memory alloy fastener rings, in: Engineering Aspects of Shape Memory Alloys, Eds. T.W. Duerig, K.N. Melton, D. Stöckel, and C.M. Wayman, Butterworth-Heinemann, London (1990), pp. 158–169

[227] T. Todoroki, Shape memory sensor and actuator for air conditioners, in: Engineering Aspects of Shape Memory Alloys, Eds. T.W. Duerig, K.N. Melton, D. Stöckel, and C.M. Wayman, Butterworth-Heinemann, London (1990), pp. 315–329

[228] E. Runtsch, Shape memory actuators in circuit breakers, in: Engineering Aspects of Shape Memory Alloys, Eds. T.W. Duerig, K.N. Melton, D. Stöckel, and C.M. Wayman, Butterworth-Heinemann, London (1990), pp. 330–337

[229] H. Horikawa, Applications for SMAs in Japan, Proc. SMST 99, Antwerpen, Belgium (1999), pp. 256–266

[230] D. Stöckel, Shape memory actuators for automotive applications, in: Engineering Aspects of Shape Memory Alloys, Eds. T.W. Duerig, K.N. Melton, D. Stöckel, and C.M. Wayman, Butterworth-Heinemann, London (1990), pp. 283–294

[231] P. Dario, R. Valleggi, M.C. Carrozza, M.C. Montesi, and M. Cocco, Microactuators for microrobots: a critical survey, J. Micromech. Microeng. 2 (1992), pp. 141–157

[232] M. Mertmann, E. Hornbogen, and K. Escher, Development of a robotic gripper based on a NiTi silicone composite material with integrated sensing functions, Proc. Int. Symposium on Shape Memory Materials, Beijing, China, Eds. Y. Chu and H. Tu, (1994), pp. 556–560

[233] Y. Suzuki, Fabrication of shape memory materials, in: Shape Memory Materials, Eds. K. Otsuka and C.M. Wayman, Cambridge University Press (1998), pp. 133–148

[234] D. Honma, Y. Miwa, and N. Iguchi, Application of shape memory effect to digital control actuator, Bulletin of JSME, Vol. 27, No. 230 (1984), pp. 1737–1742

[235] J. Hesselbach, R. Pittschellis, E. Hornbogen, and M. Mertmann, Shape memory alloys for use in miniature grippers, Proc. SMST 97, Pacific Grove, California, USA, (1997), pp. 251–256

[236] K. Kuribayashi, A new actuator of joint mechanism using TiNi alloy wire, Int. J. Robotics Res. 4 (1986), pp. 47–58

[237] K. J. Gabriel, W.S.N. Trimmer, and J.A. Walker, A micro rotary actuator using shape memory alloys, Sensors and Actuators 15 (1988), pp. 95–102

[238] K. Kuribayashi, Millimeter-sized joint actuator using a shape memory alloy, Sensors and Actuators 20 (1989), pp. 57–64

[239] K. Ikuta, M. Tsukamoto, and S. Hirose, Shape memory alloy servo actuator system with electric resistance feedback and application for active endoscope, Proc. of IEEE International Conference on Robotic and Automation (1988), pp. 427–430

[240] P. Dario, R. Valleggi, M. Perdini, and A. Sabatini, A miniature device for medical intracavitary intervention, IEEE Catalog No. CH2957-9, (1991), pp. 171–175

[241] S. Maeda, O. Tohyama, and H. Ito, Micro actuators with shape memory alloy (SMA) coil springs and their application to active fiberscopes, IEICE Trans. Electron., Vol. E80-C, No. 2 (1997), pp. 226–231

[242] K. Kuribayashi, S. Shimizu, T. Nishinohara, T. Taniguchi, M. Yoshtake, and S. Ogawa, Trial fabrication of micron sized arm using reversible TiNi alloy thin film actuators, Proc. IEEE/RSJ 93, Yokohama, Japan (1993), pp. 1697–1702

[243] M. Kohl, E. Quandt, A. Schüßler, R. Trapp, and D.M. Allen, Characterization of NiTi shape memory microdevices produced by microstructuring of etched sheets or sputter deposited films, Proc. Actuator 94, Bremen, Germany, Eds. H. Borgmann and K. Lenz, (1994), pp. 317–320

[244] W.L. Benard, H. Kahn, A.H. Heuer, and M.A. Huff, A titanium-nickel shape memory alloy actuated micropump, Proc. Transducers 97, Chicago, USA (1997), pp. 361–364

[245] S. Takeuchi and I. Shimoyama, Three dimensional SMA microelectrodes with clipping structure for insect neural recording, Proc. MEMS 99, Orlando, Florida, USA (1999)

[246] A.P. Lee, D.R. Ciarlo, P.A. Krulevitch, S. Lehew, J. Trevino, and M.A. Northrup, A practical microgripper by fine alignment, eutectic bonding and SMA actuation, Sensors and Actuators A54 (1996), pp. 755–759

[247] Y. Furuya, M. Matsumoto, H.S. Kimura, K. Aoki, and T. Masumoto, Mater. Trans. JIM 31 (1990), p. 504

[248] K. Kuhnen and H. Janocha, Compensation of the creep and hysteresis effects of piezoelectric actuators with inverse systems, Proc. Actuator 98, Bremen, Germany (1998), pp. 309–312

[249] M. Kohl, D. Dittmann, E. Quandt, B. Winzek, S. Miyazaki, and D.M. Allen, Shape memory microvalves based on thin films or rolled sheets, Proc. ICOMAT 98, San Carlos de Bariloche, Argentina (1998)

[250] A. Strojek, Entwicklung von druckkompensierten Mikromembranventilen mit NiTi-Formgedächtnisantrieb, diploma thesis, University of Karlsruhe (1998)
[251] D. Dittmann, Mikroventile mit Formgedächtnis-Dünnschichten, diploma thesis, University of Karlsruhe (1998)
[252] J. Göbes, Entwicklung von Normally-Closed-Mikroventilen mit Formgedächtnis-Dickfilmen, diploma thesis, University of Karlsruhe (1999)
[253] I. Hürst, Thermische Optimierung eines Mikroventils mit Formgedächtnisantrieb, diploma thesis, University of Karlsruhe (1999)
[254] S. Hoffmann, Formgedächtnis-Mikroventile für die Laboranalytik, diploma thesis, University of Karlsruhe (2000)
[255] K.D. Skrobanek, M. Kohl, and S. Miyazaki, Stress-optimized shape memory microvalves, Proc. MEMS 97, Nagoya, Japan, IEEE Catalog No. 97CH36021, (1997), pp. 256–261
[256] M. Kohl, E. Just, A. Strojek, K.D. Skrobanek, W. Pfleging, and S. Miyazaki, Shape memory microvalves for high pressure applications, Proc. Actuator 98, Bremen, Germany (1998), pp. 473–477
[257] M. Kohl, I. Hürst, and B. Krevet, Time response of shape memory microvalves, Proc. Actuator 2000, Bremen, Germany, Ed. H. Borgmann, (2000), pp. 212–215
[258] M. Kohl, J. Göbes, and B. Krevet, Normally-closed shape memory microvalve, Proc. SMART 2000, Sendai, Japan (2000); Int. J. of Appl. Electromagnetics and Mechanics 12 (2000), pp. 71–77
[259] M Kohl, D. Dittmann, E. Quandt, B. Winzek, S. Miyazaki, and D.M. Allen, Shape memory microvalves based on thin films or rolled sheets, Mat. Science & Eng. A273–275 (1999), pp. 784–788
[260] M. Kohl, K.D. Skrobanek, and S. Miyazaki, Development of stress-optimised shape memory microvalves, Sensors and Actuators A72 (1999), pp. 243–250
[261] W.K. Schomburg, W. Bacher, W. Bier, B. Büstgens, J. Fahrenberg, C. Goll, D. Maas, W. Menz, and D. Seidel, Fabrication of microfluidic devices by thermoplastic molding and diaphragm transfer, Proc. ASME 95, San Francisco, California, USA (1995), pp. 951–956
[262] W. Kalide, Einführung in die Technische Strömungslehre, Carl Hanser, Munich (1990)
[263] P. Gravesen, J. Branebjerg, and O.S. Jensen, Microfluidics - a review, J. Micromech. Microeng. 3 (1993), pp. 168–182
[264] Dubbel, Taschenbuch für den Maschinenbau, Eds. W.Beitz and K.H. Küttner, Springer, Berlin (1983)
[265] Th. Haas, Laserstrahl-Schweißen von NiTi-Formgedächtnislegierungen, thesis, University of Karlsruhe (1996)
[266] H. Haferkamp, S. Paschko, F. von Alvensleben, and M. Goede, Laser material processing of shape memory alloys, Proc. Actuator 2000, Bremen, Germany, Ed. H. Borgmann, (2000), pp. 163–166

[267] D.M. Allen, T. Leong, S.H. Lim, and M. Kohl, Photofabrication of the third dimension of NiTi shape memory alloy microactuators, Proc. Int. Conf. on Microlithography and Metrology in Micromachining, Austin, Texas, USA, Eds. M.T. Postek and C. Friedrich, SPIE Vol. 3225, (1997), pp. 126–132
[268] Y. Liu, S. Dürr, D. Dittmann, and M. Kohl, A modular integrated microfluidic controller, Proc. Actuator 02, Bremen, Germany, Ed. H. Borgmann, (2002), pp. 231–234
[269] M. Kohl and B. Krevet, 3D simulation of a shape memory microactuator, Materials Transactions, Vol. 43, No. 5, (2002), pp. 1030–1036
[270] U. Gellert, Siemens Inc., Karlsruhe, Germany, private communication
[271] H. Töpfer, Bürkert Inc., Großröhrsdorf, Germany, private communication
[272] E. Just, Entwicklung eines Formgedächtnismikrogreifers, thesis, University of Karlsruhe (2000)
[273] K. Hara, K. Hane, M. Sasaki, and M. Kohl, Si micromechanical fiber-optic switch with shape memory alloy microactuator, Proc. Transducers 99, Sendai, Japan (1999), pp. 790–793
[274] E. Just, M. Kohl, W. Pfleging, and S. Miyazaki, SMA microgripper with integrated antagonism, Proc. Transducers 99, Sendai, Japan (1999), pp. 1768–1771
[275] E. Just, M. Kohl, W. Pfleging, and S. Miyazaki, SMA microgripper, Proc. SMST 99, Antwerp Zoo, Belgium (1999), pp. 280–286
[276] E. Just, M. Kohl, and S. Miyazaki, Position control of a SMA microgripper, Proc. ESOMAT 2000, Como, Italy (2000)
[277] M. Kohl and K.D. Skrobanek, Linear microactuators based on the shape memory effect, Sensors and Actuators A 70 (1998), pp. 104–111
[278] M. Kohl, E. Just, W. Pfleging, and S. Miyazaki, SMA microgripper with integrated antagonism, Sensors and Actuators A83 (2000), pp. 208–213
[279] Y. Bellouard, T. Lehnert, J.-E. Bidaux, T. Sidler, R. Clavel, and R. Gotthardt, Local annealing of complex mechanical devices: a new approach for developing monolithic micro-devices, Proc. ICOMAT 98, San Carlos de Bariloche, Argentina (1998)
[280] Y. Bellouard, R. Clavel, J.-E. Bidaux, R. Gotthardt, and T. Sidler, A new concept for monolithic shape memory alloy micro-devices used in microrobotics, Proc. Actuator 98, Bremen, Germany, Ed. H. Borgmann (1998), pp. 499–502
[281] K. Kuribayashi and S. Shimizu, Direct Joule's heat crystallization treatment for TiNi thin film microactuator, Proc. Actuator 98, Bremen, Germany, Ed. H. Borgmann (1998), pp. 465–468
[282] C. Marxer, M.A. Gretillat, N.F. de Rooij, R. Battig, O. Anthamatten, B. Valk, and P. Vogel, Vertical mirrors fabricated by reactive ion etching for fiber optical switching applications, Proc. MEMS 97, Nagoya, Japan (1997), pp. 49–54

[283] H. Toshiyoshi and H. Fujita, Electrostatic microtorsion mirrors for an optical switch matrix, Journal of MEMS, Vol. 5, No. 4 (1996), pp. 231–237
[284] S. Nagaoka, Y. Suzuki, and H. Nakada, Compact wavelength selector using a single-mode fiber switch equipped with microoptical fiber chips, Proc. MOEMS 97, Nara, Japan (1997), pp. 165–170
[285] M. Hoffmann, P. Kopka, and E. Voges, Bistable micromechanical fiber-optic switches on silicon, Proc. MOEMS 98 (1998), pp. 31–35
[286] K. Hara, Mimaki Electronic Component Co., Ltd., Nagano, Japan
[287] Y. Tatsue and T. Kitahara, Micro-grip system, J. of Robotics and Mechatronics, Vol. 3, No. 1 (1990), pp. 57–59
[288] J. Hesselbach and R. Pitschellis, Greifer für die Mikromontage, wt-Produktion und Management 85, Springer Verlag Berlin (1995), pp. 595–600
[289] G. Greitmann and R.A. Buser, Tactile microgripper for automated handling of microparts, Sensors and Actuators A 53 (1996), pp. 410–415
[290] G. Thornell, M. Bexell, J.A. Schweitz, and S. Johansson, Design and fabrication of a gripping tool for micromanipulation, Sensors and Actuators A 53 (1996), pp. 428–433
[291] R. Salim and H. Wurmus, Multi gearing compliant mechanisms for piezoelectric actuated microgrippers, Proc. Actuator 98, Bremen, Germany, Ed. H. Borgmann (1998), pp. 186–188
[292] K.D. Müller, K. Pollak, W. Bacher, M. Heckele, and U. Gengenbach, Holistic concept of sensitive gripping and magazining of microcomponents, Proc. Microsystems Technologies 98, Potsdam, Germany (1998), pp. 243–248
[293] M. Kohl, B. Krevet, and E. Just, SMA microgripper system, Sensors and Actuators A97–98 (2002), pp. 646–652
[294] B. Köhler and F. Eberle, A digital image processing system to automate the production of microstructures with hot stamping technology, Proc. Microsystems Technologies 98, Potsdam, Germany (1998)
[295] U. Gengenbach, A. Hofmann, B. Köhler, E. Just, M. Kohl, F. Eberle, F. Engelhardt, and R. Scharnowell, Komponenten zum Aufbau modularer Mikrofertigungssysteme, 4. Statuskolloquium des Projekts Mikrosystemtechnik, Forschungszentrum Karlsruhe (2000), pp. 203–204
[296] R.P. Feynman, There's plenty of room at the bottom, Annual meeting of the APS at the California Institute of Technology, December 1959; reprint: J. of MEMS 1, No. 1 (1992), pp. 60–66
[297] W. Menz and J. Mohr, Das Potential der Mikrosystemtechnik in der Medizin, in: Mikrosystemtechnik für Ingenieure, VCH Verlagsgesellschaft, 2nd edition (1997)
[298] L. Fan, H. Last, R. Wood, B. Dudley, C. Khan-Malek, and Z. Ling, SLIGA based underwater weapon safety and arming system, Proc. HARMST 97, Madison, Wisconsin, USA (1997)

[299] M.S. Rodgers, J.J. Sniegowski, J.J. Allen. S.L. Miller, J.H. Smith, and P.J. McWhorter, Intricate mechanisms-on-a-chip enabled by 5-level surface micromachining, Proc. Transducers 99, Sendai, Japan (1999), pp. 990–993

Subject Index

MIX
Papier aus verantwortungsvollen Quellen
Paper from responsible sources
FSC® C105338

If you have any concerns about our products,
you can contact us on
ProductSafety@springernature.com

In case Publisher is established outside the EU,
the EU authorized representative is:
Springer Nature Customer Service Center GmbH
Europaplatz 3, 69115 Heidelberg, Germany

Printed by Libri Plureos GmbH
in Hamburg, Germany